ASME International

AN ASME REPORT

CRTD Vol. 60

Final Report

REFERENCE METHOD ACCURACY AND PRECISION (ReMAP): PHASE 1

Precision of Manual Stack Emission Measurements

Prepared by:

W. Steven Lanier
GE Energy and Environmental Research Corporation

Charles D. Hendrix
Statistical Consultant

under the auspices of:
American Society of Mechanical Engineers
Research Committee on Industrial and Municipal Waste

February 2001

Disclaimer

This report was prepared through the collaborative efforts of The American Society of Mechanical Engineers (ASME) Center for Research and Technology Development (hereinafter referred to as the Society or ASME) and sponsoring companies.

Neither the Society nor the Sponsors, nor the Society's subcontractors, nor any others involved in the preparation or review of this report, nor any of their respective employees, members, or other persons acting on their behalf, make any warranty, expressed or implied, or assume any legal liability or responsibility for the accuracy, completeness, or usefulness of any information, apparatus, product, or process disclosed or referred to in this report, or represent that any use thereof would not infringe privately owned rights.

Reference herein to any specific commercial product, process, or service by trade name, trademark, manufacturer, or otherwise does not necessarily constitute or imply its endorsement, recommendation, or favoring by the Society, the Sponsor, or others involved in the preparation or review of this report, or agency thereof. The views and opinions of the authors, contributors, and reviewers of the report expressed herein do not necessarily reflect those of the Society, the Sponsors, or others involved in the preparation or review of this report, or any agency thereof.

Statement from the by-laws of the Society: The Society shall not be responsible for statements or opinions advanced in its papers or printed in publications (7.1.3).

Authorization to photocopy material for internal or personal use under circumstances not falling within the fair use provisions of the Copyright Act is granted by ASME to libraries and other users registered with the Copyright Clearance Center (CCC) provided that the applicable fee is paid directly to the CCC, 222 Rosewood Drive, Danvers, MA 01923. [Telephone: (978) 750-8400] Requests for special permission or bulk reproduction should be addressed to the ASME Technical Publishing Department.

Reference Method Accuracy and Precision (ReMAP Phase I)
An Assessment of the Precision of
EPA Manual Stack Emission Measurements

Executive Summary

This report documents results from the first phase of a study co-sponsored by the American Society of Mechanical Engineers (ASME) to assess the accuracy and precision of manual test methods adopted by the US Environmental Protection Agency (EPA) for determining the stack concentration of selected air pollutants. The program is entitled Reference Method Accuracy and Precision and is referred to by the acronym ReMAP. The Phase 1 effort addresses the precision of the selected measurement methods. The formal Purpose Statement for the program is:

> "To determine the precision of pollutant emission measurements based on analysis of available simultaneous – sample test data which were generated using EPA Manual Reference Test Methods 5 and 5i (Particulate Matter), 23 (Dioxin and Furan), 26 (HCl), 29 (multi-metals), 101a and 101b (mercury) and 108 (arsenic) at a number of stationary air sources."

As used in the ReMAP program, precision is defined as random error that inadvertently enters the measurement process. This error may enter at any stage of the measurement process including sample collection, sample recovery, or sample analysis. The impact of such errors is that measurement results deviate from the true stack concentration. Because these errors occur randomly, data from repeated application of any manual test method (to a hypothetical stack with a chosen average concentration of the selected pollutant) should result in a bell shaped frequency distribution. This distribution is centered about the chosen stack concentration but data are expected indicating measured concentrations both above and below the chosen stack value.

Precision of a measurement method is indicated by the horizontal spread of the bell curve. One common way to characterize the bell curve shape is to determine the estimated standard deviation (σ) of the distribution. Alternately, the bell distribution also describes the range of measurement results anticipated from repeated application of the measurement method at the given concentration. Based upon the best estimate of the standard deviation, it is a simple matter to calculate other indicators of data quality that have direct practical significance. The ReMAP program selected two additional, directly derived parameters to characterize precision of manual methods:

1. The anticipated range for 99 out of 100 future single measurements. If the measurement method is repeatedly applied to a stack with a given concentration, this precision metric defines the upper and lower concentration bounds for 99% of individual measurements.

2. The anticipated range for 99 out of 100 future triplicate measurements. Since most environmental regulations define the reportable stack concentration as the average of three repeated test runs, this metric defines the anticipated range of results in triplicate (3 single time series) measurements due to random error in the measurement process.

For each of these precision metrics it is important to note the inherent assumption that the measurement method is being applied to a hypothetical stack with an average pollutant concentration that does not vary with time. In real-world applications, single and triplicate measurement results will indicate variation from both random errors and systematic errors (bias) in the measurement process as well as temporal variation in facility operation.

As noted, the primary objective of the ReMAP program is to characterize the bell shaped curve for each measurement method by first estimating the standard deviation σ. The other two precision metrics are directly calculated from σ. Specifically 99 out of 100 single measurements will fall within $\pm 2.54\ \sigma$ of the true concentration. The anticipated range for the average of repeated measurements comes progressively closer to the true concentration. More precisely, the anticipated range varies inversely with the square root of the number of measurements. Thus, if σ is the standard deviation of anticipated single measurements, the anticipated range for 99 out of 100 future triplicate measurements will fall within $\pm 2.54\ \sigma\ /\sqrt{3}$ of the true concentration.

A central tenet of the ReMAP program is to determine how the precision of manual measurement methods varies with the concentration of the pollutant being measured. A constant measurement standard deviation might occur if the primary source of random error is an analytical process. A simple example might be the repeatability with which a technician can measure the weight gain on a particulate filter. The capabilities of the weighing scale do not significantly vary with the particulate weight gain. However, the magnitude of random errors associated with extraction and recovery of the sample from the stack might be expected to vary in proportion to stack concentration.

The technical effort of the ReMAP program is to estimate the standard deviation of selected EPA Measurement Reference Methods as a function of average stack concentration. ReMAP is not intended to validate any of the Reference Methods addressed herein nor is it to be used as a substitute for Method 301.

The concentration of pollutants released from industrial facilities must be assumed to vary with time. Accordingly, results from typical, single-train stack tests are of limited value for determination of the precision of a measurement method. Instead, estimation of method precision must be based on data from special tests where multiple sampling trains are used to simultaneously determine the stack pollutant concentration. Such multi-train tests minimize the impact of temporal and spatial (where probes are co-located) variations on the data and the results can be used to estimate the standard deviation of the particular measurement at the specific stack concentration.

The ReMAP program performed a careful assessment of the statistical analysis procedures required to estimate the precision of Manual Reference Methods using multi-train sampling data (see Appendix). To assure the quality of data used in the statistical analysis, an extensive effort was expended in gathering data from the original sources and carefully evaluating them to assure that consistent data reduction procedures were used.

Conceptually, the ReMAP statistical analysis procedure is straightforward. First, data from a multi-train test run are averaged to provide an estimate of the average concentration for the run (C_i). The standard deviation for the test run (S_i) is also calculated. Clearly, a calculated standard deviation from a single test using a dual sampling probe provides a relatively poor estimate of the true standard deviation of the method (σ) at the true concentration (μ). However, after accounting for various biases, a significant array of data from multi-train tests should provide a reasonable basis for estimating the true standard deviation as a function of concentration. The ReMAP procedure is to assume that the standard deviation varies with concentration according to a power function relationship and then to fit the data to that equation using regression analysis.

Results from the regression analysis represent the best estimate available on the standard deviation of the measurement method at any given concentration. However, the ReMAP analysis procedure also provides for calculation of confidence intervals on the regression. These confidence intervals define the upper and lower bounds for the regression line at the 95% confidence level. Based on the regression line and the confidence intervals, the various precision metrics can be determined for each Method as a function of concentration. Results from the analyses are summarized below.

EPA Method 5 and 5i for Particulate Matter – Front Half Only
Method 5 was one of the first EPA Reference Methods developed for stack sampling. There is a relatively large body of multi-train data indicating that the relative standard deviation (RSD) has minimal variation with concentration. Over a broad concentration range, RSD is predicted to remain between 5 and 11%. There are certain data comparability concerns associated with the Method 5 precision analysis. However, assuming that the Method is applied to a stack with a particulate matter concentration less than 150 mg/dscm, the best available estimate is that RSD will be below 10%. Moreover, the influence of random error in the measurement process should result in the average of triplicate measurements deviating from the true average concentration by no more than about ±14.7% (±2.54*10%/√3).

Method 5i was specifically developed for application to stacks with particulate concentrations below 50 mg/dscm. The Method itself requires dual-train sampling and provides an upper limit on the allowable deviation between the simultaneous measurements. Thus the available data had been prescreened to eliminate test results with large standard deviation (i.e., > about 14% RSD). This prescreening, coupled with the relatively small concentration range for the data, resulted in the ReMAP analysis finding no statistically significant variation of standard deviation with concentration.

Based on a pooled analysis, the characteristic standard deviation for Method 5i was found to be 1.43 mg/dscm. Based on this best estimate of standard deviation, the ReMAP analysis indicates that 99 out of 100 Method 5i single measurements should deviate from the true concentration by no more than ± 3.68 mg/dscm. For triplicate measurements 99 out of 100 Method 5i data results should deviate from the true concentration by no more than ± 2.12 mg/dscm.

Method 23 for Dioxin and Furan

Data collected with Method 23 are used to report dioxin and furan emissions as either the total mass of tetra through octa chlorinated dioxin plus furan or as the toxic equivalent emission, adjusting the mass of each congener according to its toxicity relative to 2,3,7,8 TCDD. The individual 2,3,7,8 substituted dioxin and furan congeners in tetra through octa homologues are weighted by specific factors ranging from zero to 1.0 to determine the toxic equivalent of 2,3,7,8 TCDD.

The precision of Method 23 was assessed when data were reported in both forms. Even though the same data were used for both assessments, the ReMAP results suggest that the measurement precision varies according to how the data are reported. This implies that the random error associated with the sum of all the congener masses is different than random errors associated with the sum of weighted masses.

The ReMAP assessment of Method 23 was performed using a limited database of multi-train emissions data. For application of Method 23 for determination of total PCDD/PCDF mass, the ReMAP analysis found that RSD varied between about 6.3% and 20% for stack concentrations in the range of 2 to 27 ng/dscm. The following table presents the anticipated upper and lower bounds for 99 out of 100 Method 23 measurements as a function of the true stack concentration.

Table ES-1. Anticipated Range of Measurement Results Due to Random Error in Application of Method 23 for Total PCDD/PCDF Determination.

True Stack Concentration ng/dscm	99 out of 100 Single Measurements		99 out of 100 Triplicate Measurements	
	Lower Limit	Upper Limit	Lower Limit	Upper Limit
2	0.97	3.03	1.40	2.60
6	4/09	7.91	4.90	7.10
10	7.46	12.5	8.53	11.5
14	10.9	17.1	12.2	15.8
18	14.5	21.5	16.6	20.0
22	18.1	25.9	19.7	24.3
26	21.7	30.3	23.5	28.5

As noted, indicated measurement precision is different when Method 23 is used to determine concentration on a toxic equivalence basis. Specifically, the regression analysis found no statistically significant variation of standard deviation with concentration. Pooled analysis indicates that the best estimate of standard deviation is 0.027 ng ITEQ/dscm. when the emission concentration is in the range of 0.02 to 0.9 ng ITEQ/dscm. This further indicates that 99 out of 100 future single measurements should fall with ±0.069 ng ITEQ/dscm of the true concentration and 99 out of 100 triplicate measurements should fall within ±0.04 ng ITEQ/dscm of the true concentration.

The absolute value of anticipated range for future Method 23 measurements (as ITEQ) are quite small in absolute terms but they are on the same order as regulatory emission limits being considered in some regions. As indicated above, the best estimate of standard deviation is 0.027 ng ITEQ/dscm. However, at 95% confidence, the standard deviation may be as large as 0.037 ng ITEQ/dscm and the potential range for 99 out of 100 future measurements might deviate from the true concentration by as much as ±0.095 ng ITEQ/dscm. Relying upon a single measurement has the potential to create problematic findings. If emission limits were set at 0.095 ng ITEQ/dscm. to be assured of compliance at the 95% confidence level, measurement results could not exceed zero. Similarly, measurement results must be above 0.19 ng ITEQ/dscm to establish, with 95% confidence that the true stack concentration exceeded the emission limit.

Most regulations and permit limits establish compliance based on averaging results from triplicate measurements. The anticipated range for 99 out of 100 future triplicate measurements is reduced, relative to single measurements, by √3. Thus, compliance with an emission limit of 0.095 ng ITEQ/dscm is assured (at the 95% confidence level) when the triplicate average is at or below 0.04 ng ITEQ/dscm. Similarly, at 95% confidence, exceedence of the 0.095 ng ITEQ/dscm limit is assured when the three run average is above 0.15 ng ITEQ/dscm.

Method 26 for Hydrochloric Acid

ReMAP analysis of available data for Method 26 for HCl indicated that RSD is typically in the range of 5% to 10%. RSD does increase when the method is applied to stacks with very low concentration. Table ES-2 summarized the anticipated upper and lower bounds for 99 out of 100 Method 26 measurements as a function of true stack HCl concentration.

Table ES-2. Anticipated Range of HCl Measurement Results Due to Random Error in Application of Method 26.

True Stack HCl Concentration mg/dscm	99 out of 100 Single Measurements		99 out of 100 Triplicate Measurements	
	Lower Limit	Upper Limit	Lower Limit	Upper Limit
1	0.65	1.35	0.80	1.20
5	3.72	6.28	4.26	5.74
10	7.76	12.2	8.71	11.3
20	16.1	23.9	17.7	22.3
50	41.9	58.1	45.3	54.7
100	85.8	114.2	91.8	108.2

Methods 29, 101a and 101b for Total Mercury

Several measurement methods have been developed for measurement of total emission concentration and for mercury speciation. The ReMAP analysis took all available multi-train mercury data collected using Methods 29, 101a and 101b but only used the data for total mercury concentration. The data analysis indicates that over the concentration range of 50 to 783 μg/dscm, the measurement method RSD varied from 9.6 to 12.4%. As concentration drops from 50 to 5 μg/dscm, the RSD is expected to rise from 12.4% to 15.4%. Table ES-3 summarizes the anticipated upper and lower bounds for 99 out of 100 mercury measurements using Methods 29 and 101 as a function of true stack total mercury concentration.

Table ES-3. Anticipated Range of Total Hg Measurement Results Due to Random Error in Application of Methods 29, 101a and 101b.

True Stack Hg Concentration μg/dscm	99 out of 100 Single Measurements		99 out of 100 Triplicate Measurements	
	Lower Limit	Upper Limit	Lower Limit	Upper Limit
4	2.19	5.81	2.96	5.04
10	5.96	14.0	7.70	12.3
25	16.0	34.0	19.8	30.2
50	33.4	66.6	40.4	59.6
75	51.3	98.7	61.7	88.7
100	69.5	130.5	82.4	117.6

Method 29 for Multi-Metals

Method 29 is also used for measurement of several other metal emissions. Precision analysis was completed for six other metals including antimony, arsenic, beryllium, cadmium, chromium, and lead. With the exception of cadmium, the analysis indicates that these metals behave similarly with respect to measurement method precision. A composite analysis was performed for the five similarly behaving metals and the results indicate that use of Method 29 provides an RSD that varies between 13 and 18% when the individual metal concentrations are between about 20 and 100 μg/dscm. Table ES-4 summarizes the anticipated upper and lower bounds for 99 out of 100 Sb, As, Be, Cr, and Pb measurements using Methods 29 as a function of true stack total metal concentration.

As regards cadmium measurements using Method 29, the analysis indicates that standard deviation is a weaker function of concentration, at least at higher concentration ranges. The best estimate of RSD is 9.1 % when cadmium concentration is 80 μg/dscm and 18.7 % when the concentration drops to 20 μg/dscm. However, at 5 μg/dscm, predicted RSD is 38.6% and at 1.4 μg/dscm RSD is predicted to exceed 75%.

Table ES-4. Anticipated Range of Sb, As, Be, Cr, and Pb Measurement Results Due to Random Error in Application of Methods 29 (Composite Analysis).

True Stack Sb, As, Be, Cr, or PB Concentration μg/dscm	99 out of 100 Single Measurements		99 out of 100 Triplicate Measurements	
	Lower Limit	Upper Limit	Lower Limit	Upper Limit
5	2.16	7.84	3.36	6.64
20	11.0	29.0	14.8	25.2
40	24.0	56.0	30.8	49.2
60	37.6	82.4	47.1	72.9
80	51.5	108.5	63.6	96.4
100	65.7	134.3	80.2	119.8

REFERENCE METHOD ACCURACY AND PRECISION (ReMAP)
PHASE I
PRECISION OF MANUAL STACK EMISSION MEASUREMENTS

1.0 Introduction

An integral part of efforts to regulate and control air pollution emissions is collection and analysis of exhaust stream samples to determine the concentration and flow rate of pollutants released to the atmosphere. The U.S. Environmental Protection Agency (EPA) and its counterparts in other countries have developed formal methods defining the hardware and procedures for collecting and analyzing samples to quantify emissions of individual pollutants. A significant number of these methods involve manual extraction of a sample from a facility's exhaust stack, sample recovery and subsequent laboratory analysis to quantify concentration of a specific pollutant(s) in the sample. All manual processes, including the various EPA measurement methods, are subject to random variations, which ultimately impact the end results. Relatively minor variations in the skill of the sampler, as well as the equipment and procedures used to extract the sample can influence the indicated quantity of sample extracted from the stack and the efficiency with which the pollutant of interest is collected or recovered. Similarly, minor variation in laboratory hardware and procedures influence quantification of the mass or volume of pollutant in that sample. The net result of such random variation is imprecision in measurement results. The current report documents a study where available data have been gathered and analyzed to quantify the precision of key EPA manual measurement methods. The study has been conducted under the auspices of the American Society of Mechanical Engineers (ASME) and is entitled **Reference Method Accuracy and Precision**[1] **(ReMAP), Phase 1.**

The purpose of the ReMAP – Phase 1 program is "*to determine the precision of pollutant emission measurements based on analysis of available simultaneous-sample*[2] *test data which were generated*

---

[1] Precision is defined here as "Random Error" according to the new ASME PTC 19.1-1998.
[2] Dual-train, quad-train, and simultaneous-samples from different sample locations at a stationary emission source.

1

using EPA Manual Reference Test Methods 5 and 5i (PM), 23 (dioxin and furan), 26 (HCl), 29 (multi-metals, 101a and 101b (mercury), and 108 (arsenic) at a number of stationary air sources." ASME intends ReMAP to be a multi-phase effort with the first phase focusing exclusively on assessment of measurement method precision. Consideration of issues associated with measurement accuracy is reserved for a later phase of ReMAP.

Three major groups have sponsored the ReMAP-Phase 1 effort. First, the U.S. EPA has provided funding and personnel support to the project. Second, several industrial groups representing manufacturing companies and the waste combustion industry have provided program funding. Finally, the ASME's Committee on Industrial and Municipal Waste has provided both financial support and overall program direction.

Although Phase I results indicate that the various EPA Methods provide differing levels of precision and that the precision typically varies with stack pollutant concentration. ReMAP does not reach conclusions relative to policy issues such as how results of the study should be used. Answering those questions is appropriately left to lawmakers, regulatory authorities, regulated industries, and the public. The role of ReMAP is to provide scientifically sound data analysis and results to facilitate meaningful policy debate and decision making.

It is important to note from the outset that a variety of factors contribute to variability in measured stack emission concentrations. In addition to measurement method precision (which includes the skill of the stack tester), variation of process feed materials (including combustion fuels) and unit operations impact stack emissions. A compliance test with three successive single samples will be impacted by process variation over time and those measurements of stack emission concentrations will potentially indicate greater variability than suggested by the precision of the measurement method itself.

2.0   Background

The US EPA has developed and published a wide variety of methods for determining the concentration of pollutants in process effluent streams. The manual air sampling methods typically involve a probe for extracting a representative sample of stack effluent and means for physically capturing or chemically extracting selected pollutants from that sample. The methods further define procedures for determining the volume of sample gas extracted and for recovering the collected pollutant(s) from the sampling apparatus. Finally, the methods specify laboratory procedures to use for determining the quantity of pollutant collected.

In developing new measurement procedures, EPA has traditionally conducted extensive laboratory and field validation studies including tests to define the precision and biases of the method. Procedures employed by EPA have evolved over the years but generally conform to those described in a 1977 paper entitled, "How EPA Validates NSPS Methodology" (Midgett, 1977). Most of the procedures discussed by Midgett have been incorporated into EPA Method 301 which "is used whenever a source owner or operator proposes a test method to meet a U.S. Environmental Protection Agency (EPA) requirement in the absence of a validated method."(EPA, 1992)   In validation of a new method or in tests to evaluate an alternate method, EPA suggests use of four sampling trains to simultaneously extract samples from nominally the same location in a source stack. This is commonly referred to as a quad-train. For method validation, two of the four trains are configured and operated in strict accordance with the proposed method, while the other two trains are spiked with known quantities of the target analyte. Comparison of data from the two unspiked trains provides an indication of measurement precision while data from the spiked trains provides an indication of measurement bias. Data from a significant number of repeated multi-train runs provide an indication of the precision and bias of the method itself. Method 301 states that "The precision of the method at the level of the standard shall not be greater than 50 percent relative standard deviation."

Several of the EPA measurement methods were developed and validated in the early days of the Agency. EPA Method 5 for measuring stack particulate concentration was published in the Federal Register on December 23, 1971 (36FR 25876). Tests to validate that method were performed on sources with particulate emissions ranging from 45 to 240 mg/dscm. In that time period, the majority of Federal particulate emission standards were established at 180 mg/dscm (0.08 gr/dscf) corrected to 7% O_2. Thus, the method was validated over a range that included the prevailing regulatory limits. At this emission limit, the EPA validation studies indicate that precision of Method 5, expressed, as a relative standard deviation was on the order of 10%.

Passage of the Clean Air Act Amendments of 1990 ushered in a new era in both scope and stringency of environmental regulations. Rules governing release of Hazardous Air Pollutants (HAPs) called for regulation of 179 specific pollutants from both new and existing emission sources. Provisions in the law, stipulate that standards must consider the "Maximum Achievable Control Technology" (MACT). For existing sources, MACT standards shall not be less stringent than the average emission performance achieved by the best performing 12% of the sources in a category or subcategory. Implementation of this congressional mandate has resulted in many new emission regulations that are dramatically more stringent. For example, in 1999 the particulate emission limit for hazardous waste incinerators was tightened from 180 to 34 mg/dscm (@ 7% O_2).

Stringent emission standards raise numerous concerns about the precision of EPA measurement methods. The particulate standards can be used to highlight some of those issues. As noted, initial method validation studies assessed Method 5 precision and the Agency deemed the Method to be acceptably precise over a rather broad range of concentrations. However, that method is now being applied and results used for regulatory purposes at concentration levels significantly below the range for which it was validated. Method 5 may be acceptably precise in the lower range (e.g., <40 mg/dscm) but, prior to ReMAP Phase I, the absence of an updated method precision assessment, meant all environmental stakeholders were faced with excessive uncertainty on this issue. ReMAP brings more information to light regarding precision, however, the reader is cautioned to be aware that it is not intended to be used to validate and Reference measurement methods or as a substitute for Method 301.

Particulate matter is not the only EPA Reference measurement Method for which there is concern. Another example is EPA Method 23 for determining dioxin and furan emission concentration. The published method validation studies concentrated almost exclusively on precision and bias of

4

analytical procedures and largely ignored the sample collection portion of the overall method. No Agency method validation data were provided examining the precision of the entire sampling and analysis procedure. Method 23 does provide for extensive spiking of the sampling train with labeled compounds and includes tests to quantify recovery of those standards. However, data are considered acceptable if the fractional recovery of the labeled compounds falls within the range of 40% to 130%. With such broad allowable recoveries and in the absence of full system precision analysis, it is anticipated that Method 23 may provide results with exceedingly wide precision bands.

The above noted issues do not imply that Methods 5 and 23 are technically unacceptable procedures. Instead, these issues are typical of general concerns that develop when method validations are incomplete or out of date. In the absence of well-documented assessments of measurement method precision, many reasonable questions are formed and nurtured. Typical questions include:

- In light of the economic and public-perception consequences associated with a failed compliance test, are the current EPA measurement methods technically acceptable procedures for determining compliance with standards that have become more stringent over time?
- If a method is highly imprecise, will indication of a failed compliance test withstand scrutiny of a legal challenge?
- Do data indicating emission concentrations below the regulatory limit really imply compliance with the rules?
- Databases used to establish MACT standards are generally developed based on reports from tests using published EPA methods. A critical portion of these data - data from facilities defining the best performing 12% of the facilities - is extracted to define the MACT technology or the MACT based emission limits. Do these data characterize exceptionally well designed and operated facilities or do the key data represent imprecision in the measurement methods? This concern applies to any analysis where the best 12% of the data are selected for examination but it is even more critical when those data indicate results below the range for which the method was validated.

- Concerns extend beyond regulatory compliance and regulatory development. Are the EPA methods acceptable procedures for determining whether a new air pollution control device is meeting its performance guarantees? Is the indicated performance representative of the control device or do the test results reflect significant imprecision in the method?

The above lists of issues and concerns are far from exhaustive. However, almost invariably, the response to such questions is that the measurement methods may not be perfect but they are the best that we have. That answer does not, however, alleviate stakeholder fears. The sponsors of ReMAP, including the US EPA, have entered into the program to provide tools that might be used to develop better answers. This report is, however, not intended to be used to validate any of the measurement methods addressed herein or as a substitute for Method 301.

APPROACH

There is often confusion concerning the terms, precision and accuracy. Figure 1, adapted from a presentation by Dr. Greg Rigo at a meeting of ASME's Committee on Industrial and Municipal Waste clearly illustrates what the two terms imply. Imagine shooting at a target. The illustration on the left shows a wide scattering of results, almost equally distributed around the bull's eye. The illustration on the right shows a tightly grouped set of shots that completely miss the target. Scatter in these results is an indication of precision while proximity to the bull's eye is a measure of accuracy. The target on the left illustrates poor precision but good accuracy. The target on the right illustrates highly precise, but inaccurate shooting. Phase I of ReMAP is concerned with precision. Accuracy is an issue for later phases of ReMAP.

Many facilities have large quantities of data from repeated single-train stack tests. These results are important to the facility but typically, they shed little light on the precision of EPA measurement Methods. Variation in repeat measurements is influenced by measurement method precision but it is also influenced by variations in facility operation. Unfortunately, there are no unambiguous means for separating these two effects. Determination of measurement method precision must be based on simultaneous determinations of stack emission concentrations, preferably with co-located probes.

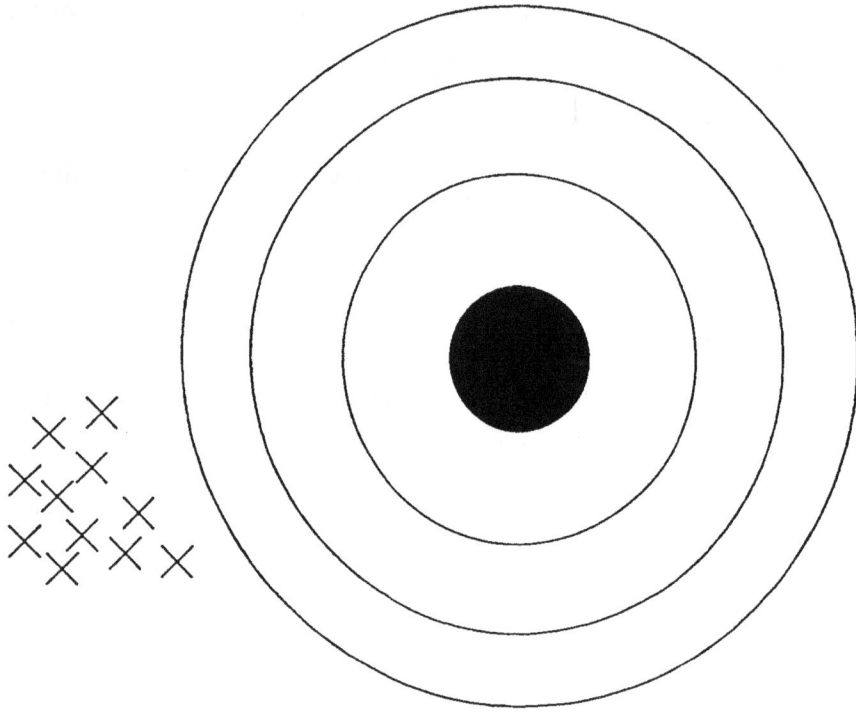

Precise But
Not Accurate

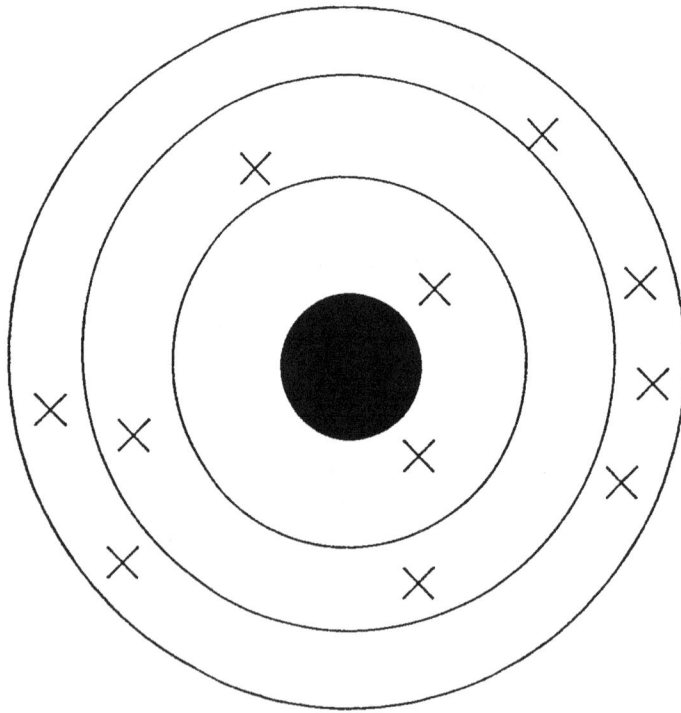

Accurate But
Imprecise

Figure 1. Illustration – The Difference Between Precision and Accuracy.

Results from two or more simultaneous measurements provide information to calculate a sample standard deviation of that measurement. Two data points – a single indication of standard deviation - are not, however, a sufficient basis for defining the precision of the measurement method. Repeated simultaneous data from a given source provide an improved indication of method precision. Repeated simultaneous measurements from a variety of facilities further improves the data base for assessing method precision since the data cover a broader range of stack concentrations and a broader range of personnel applying the method.

The ReMAP program has assembled a database consisting of available multi-train data from a variety of selected EPA methods. A key source of these data are published and unpublished EPA reports addressing method validation. Additionally, a limited number of industry sponsored, multi-train studies have been conducted and documented. These industry reports provide significant expansion to the scope of available multi-train data. Finally, the EPA has sponsored a limited number of studies where multi-train tests were performed to expand the range of data for validation of previously published methods. The ReMAP program has gathered available multi-train data sets for the following EPA measurement methods:

- Methods 5 and 5i for particulate matter (PM) emissions
- Method 23 for dioxin and furan emissions
- Method 26 for hydrochloric acid and chlorine gas
- Method 29 for multi-metals, and
- Methods 101a and 101b for mercury.

A search was made for validation data on the following EPA methods but no multi-train results were discovered:

- Method 108 for arsenic
- Method 0030 and 0010 for volatile and semi-volatile organics respectively, and
- Method 0011 for formaldehyde.

After the initial data collection activity, the ReMap program took two parallel paths. One path provided for detailed validation of the gathered data. Wherever possible, validation began with the original field run sheets and continued through a complete re-reduction of the data. This tedious process improved the database by providing consistency in such key factors as use of consistent standard reference conditions and blank correction procedures. The parallel effort involved identifying and refining mathematical procedures for analyzing simultaneously sampled concentration data to determine measurement precision at various appropriate concentrations. Finally, after validating the database, the selected statistical analysis procedures (see Appendix) were applied to the validated database to determine the precision of the selected EPA methods at appropriate concentrations.

A final preliminary point concerns the issue of correcting data to a fixed percentage of excess oxygen. Environmental regulations almost always set a limit on the concentration of pollutants in the stack and require that the concentration be adjusted to reflect a standard stack excess oxygen (typically 7% oxygen). For several reasons the ReMAP study does not include oxygen correction in the analysis of measurement method precision. The primary rationale is that the various chemical analyses determine the quantity of a specific analyte in the overall sample matrix. If the quantity of analyte is low, it makes little difference to the chemical analysis whether the loading is the result of effective air pollution control or if the stack has high excess air. A more pragmatic consideration comes from the available data. Several of the key EPA method validation studies failed to record the stack oxygen concentration during the tests.

The following material develops estimates of measurement method precision as a function of the average pollutant concentration. Both the concentration and the precision metrics (when expressed in concentration terms) can be adjusted to a fixed oxygen level by applying the following O2 correction equation:

$$[C]_{@reference O_2} = [C]_{@stack O_2} \times \frac{[20.9 - O_{2 reference}]}{[20.9 - O_{2 stack}]}$$

This, of course, requires that one have knowledge of the actual stack oxygen level as well as the desired reference oxygen level. More specifically, the precision of a measurement method, referenced to a fixed percent excess air (say 7% O_2) will vary with the O_2 concentration in the stack. This issue will be discussed in further detail in later portions of the report

Report Organization

This report is divided into two different sections. The front portion of the report has been written for readers with only a passing familiarity with statistical analysis. Included are descriptions of the measurement methods and the database of multi-train results. It also includes a layman's presentation of the data analysis procedures and a presentation of the study results. The last portion of the report (actually an appendix) provides a detailed description of the statistical analysis procedures used in ReMAP. A serious attempt has been made to make the main body of the report and the appendix readable and understandable to non-statisticians.

3.0   The Analysis Procedure - A Layman's Description

Material presented in this section provides a brief summary of statistical analysis procedures used in ReMAP to assess precision of the selected EPA measurement methods. The presentation is written for the statistical layman and may seem overly simplistic to those skilled in the statistical sciences. Such readers are referred to the report's Appendix which includes detailed development of the statistical analysis procedures.

3.1   Measures of Precision

As indicated earlier, imprecision in a measurement method implies that random error in the sampling and/or laboratory analysis result in random variation in the indicated emission concentration. Consider a hypothetical stack that emits a nearly constant concentration of some pollutant. Imprecision from the measurement method will result in measured concentrations deviating from the true stack concentration. If the hypothetical stack is sampled many times, a plot of the results should produce a bell shaped curve such as the one illustrated in Figure 2. The average of a large number of measurements should approach the true concentration and most of the data points will be relatively near the true concentration. If the average does not approach the true concentration, the measurement process is biased. Individual measurements that are significantly removed from the mean should occur with decreasing frequency. The core objective of the ReMAP program is characterization of the spread in anticipated results for different manual measurement methods and determination of how that spread varies with the stack concentration.

There are a variety of parameters that may be used to characterize the precision of a method. The U.S. EPA has historically used standard deviation or relative standard deviation to define precision. There are, however, several other parameters that may be equally valid precision indicators. In Figure 2, standard deviation (denoted by the symbol σ) is indicated as a distance on either side of the mean value in the distribution. The area under the bell shaped curve bounded by $\pm \sigma$ has special mathematical significance but for current purposes it is sufficient to note that this area covers 68.2 percent of the total area under the curve. In the example discussed

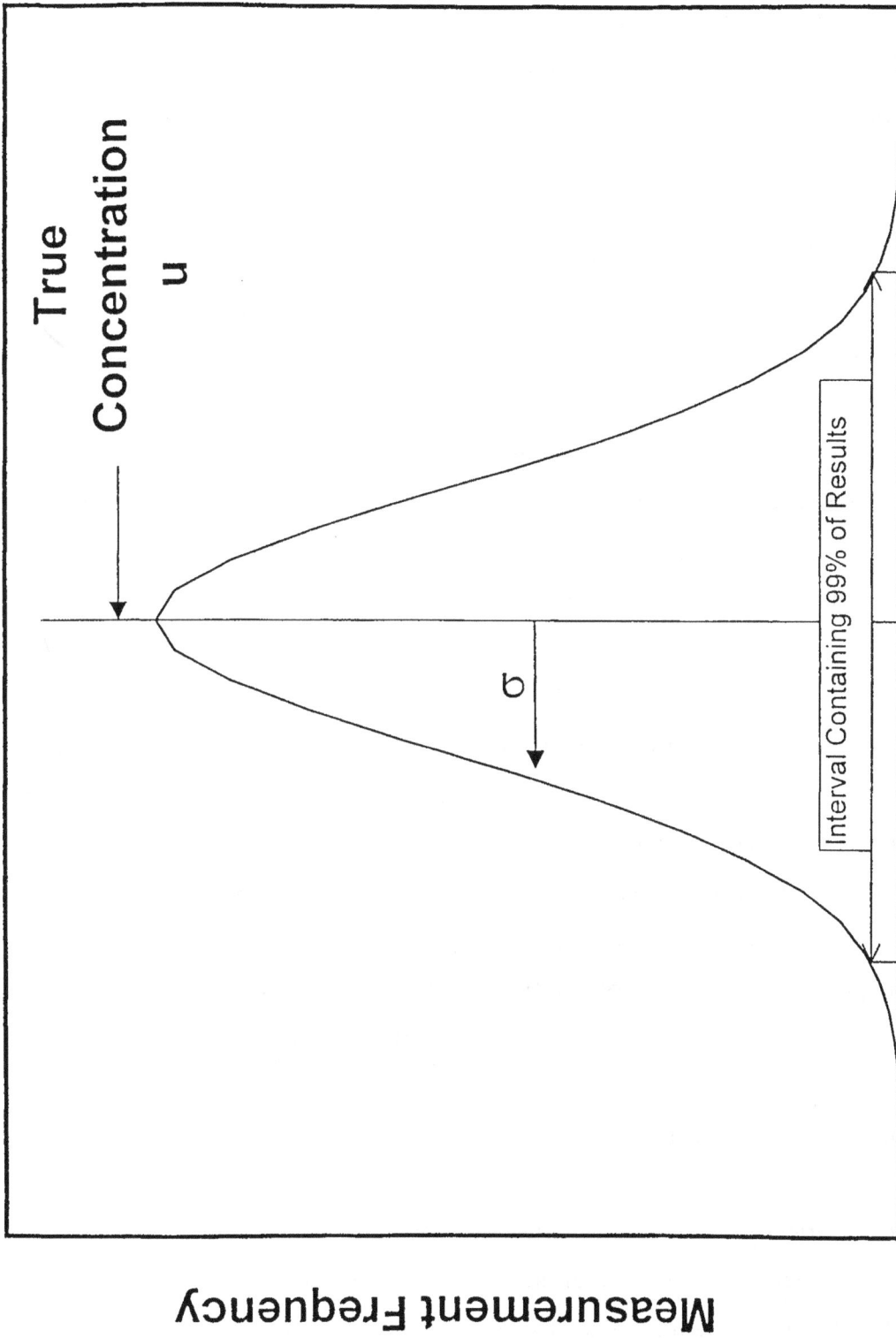

Figure 2. Normal Distribution of Field Results

above relative to Figure 2, the bell shaped curve represents the expected frequency of many future measurements. Sixty eight percent of those future measurements are expected to indicate concentrations within ± one standard deviation of the mean. If the value of σ is small relative to the mean, then the majority of the measurement results will occur close to the true value. Conversely, if the relative value of σ is large, a larger portion of the measurement results will deviate significantly from the true concentration.

A second approach for describing measurement precision is to define a data spread expected to encompass the majority of future measurements. A convenient approach is to define concentration bounds capturing 99% of the future data. This parameter may hold special significance to facility owners and operators who must comply with emission regulations. Essentially, this parameter defines the expected concentration bounds for 99 out of 100 future measurements. Where as 68.2 % of measurements fall within ± 1.0 σ of the mean, 99% of the measurements fall within ± 2.567 σ of the mean. Both the standard deviation and 99% concentration bounds represent the spread in future measurements from random variation of the measurement method. The standard deviation and 99% concentration bounds do not include variation of the emission source.

A third precision metric is the expected spread in the average of triplicate measurements. That parameter may have special significance to facility owners, since compliance with emission regulations is typically based on the average of three runs. All three precision metrics are related by simple proportion. As noted above, 99 out of 100 future single measurements will fall within the bounds of ± 2.567σ. For repeated measurements, the range decreases inversely with the square root of the number of repeat measurements. Thus, ninety-nine percent of the average of future triplicate measurements will fall within the bounds of ± 1.482 σ (± 2.567/√3 σ). All three metrics will be calculated in the ReMAP precision assessments for each measurement method.

3.2   Estimating Standard Deviation

Having settled on metrics for describing measurement precision, the problem is reduced to determining standard deviation as a function of measurement concentration. From the outset, it is critical to understand that there is a true value of standard deviation at any given concentration but we will never know its exact value. In accordance with normal statistical nomenclature, the true value of standard deviation is given the symbol σ (sigma). Information concerning σ can be obtained from special tests using two or more sampling trains operated simultaneously in the same exhaust stream. Ideally, such tests extract sample from the same nominal position(s) in the stack. Each measurement will be subject to random error, resulting in different results from the simultaneous measurements. Referring to the bell shaped distribution curve in Figure 2; these two measurements represent two data points, randomly selected from the total population of potential data points. If a large number of simultaneous measurements are taken, the individual data points should generate the full distribution. Typically, however, only two, three or four sampling trains are operated simultaneously. An estimate of the standard deviation for the measurement method is obtained by calculating the standard deviation from dual or quad-train tests according to equation 1 below.

$$ S^2 = \frac{\sum (X_i - \overline{X})^2}{(N-1)} \qquad \text{Eq. 1} $$

Standard deviation calculated from experimental data is referred to by the symbol S to make a distinction between this value and the true standard deviation σ. Clearly, selecting two random points from the full population of points that make up a bell shaped distribution provides a poor estimate of σ. Selecting four points provides a better estimate but it is also subject to considerable uncertainty. In validating measurement methods, the typical procedure is to perform repeated multi-train tests. The repeated tests provide repeated estimates of standard deviation under nearly constant concentration conditions. These data provide a reasonable basis for assessing method standard deviation at the characteristic concentration of the tests, unless the method precision is a very strong function of concentration. Such assessments provide estimated values of the true standard deviation that are referred to in this report as Est. σ.

There are however, several complications to the process of estimating the standard deviation from dual and quad-train test results. When standard deviation is calculated from small samples of a large population, the result is a biased estimate. The magnitude of the bias is dependant upon the number of data points used to estimate each value of S. A detailed presentation on the source of this bias is

14

beyond the scope of the current discussion. However, for illustrative purposes, consider the case where a large number of data pairs are randomly selected from a known distribution. Standard deviation (S) can be calculated for each data pair according to equation 1. Most non-statisticians will anticipate that the average of the standard deviations (S) calculated from many data pairs would closely approximate the true standard deviation (σ) for the overall distribution. As discussion in the Appendix, that anticipation would not be realized. In fact, using data pairs, the average standard deviation would be biased low by a factor of 1.253. If we repeated this example using three or four data points for each standard deviation calculation, the average standard deviation will be closer to the true value but the bias will still be present. For triplicate measurements, the bias factor is 1.128 while for quad-train the bias factor is only 1.085. In the ReMAP data analysis, the standard deviation calculated from each multi-train test must be multiplied by the appropriate small-sample bias-correction factor. This provides an unbiased estimation of standard deviation at the selected concentration. This calculated parameter is referred to as small-sample, bias-corrected S.

To assess the impact of pollutant concentration on method precision, it is necessary to gather multi-train measurement data over a broad range of concentration and to fit the data to an equation relating S to average concentration (C). The first step is to check these data for outliers and to prepare the data for analysis. Two approaches are used by ReMAP to screen data for outliers. These procedures are defined in the Appendix and illustrated in the next section of the report. The first procedure is known as the Dixon's-r test. This procedure is used to examine a group of data points collected during a single, multi-train test and to determine if one or more data points in the group are outliers. It is only applicable to tests with three or more simultaneous sampling trains. The second screening procedure is taken from Statistical Process Control (SPC) methods and is used to identify outliers from multiple simultaneous measurements. The essence of this procedure is to compare the span between simultaneous measurements against the weighted average span for other data in a similar concentration range. Special provisions are included to account for the fact that data may exist as pair, triplicates, or as quad-train results.

After outlier screening, the validated data are entered into a spreadsheet; standard deviations are calculated and then corrected for the above noted small sample bias. Figure 3 illustrates a hypothetical set of data showing small sample bias corrected standard deviation versus mean

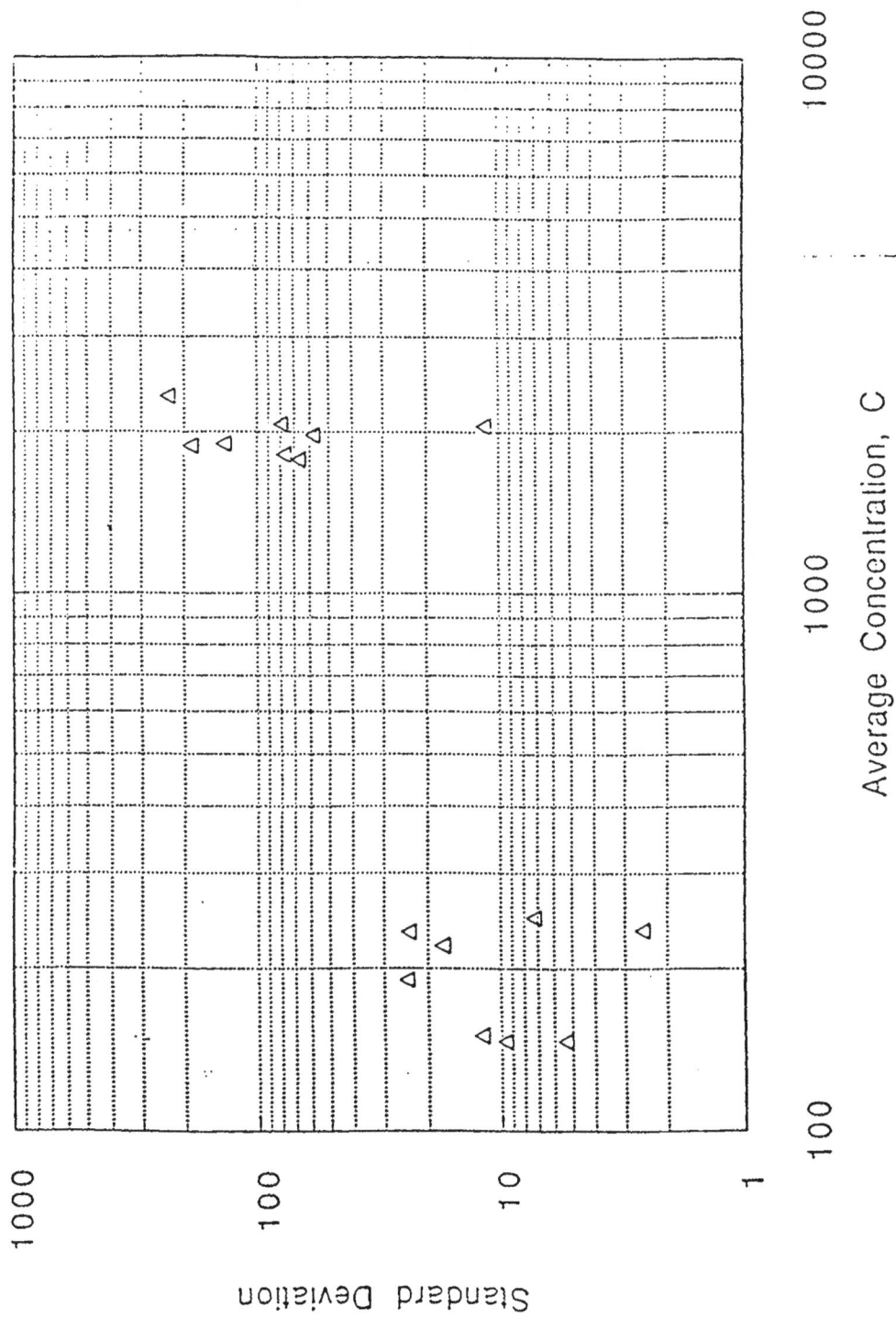

Figure 3   Simulated Data

16

concentration from several multi-train tests. The objective is to fit these data to an equation relating standard deviation ($S_{bias\ corrected}$) to average concentration (C). If done properly, this curve fit will also approximately describe the relationship between σ and C. However, before fitting the data to a functional form, it is necessary to account for the differences between data sets consisting of 2, 3, or 4 simultaneous measurements. Equation 1 can be applied to dual train measurements to determine the standard deviation (S) with only one degree of freedom. Values of S can be determined from triplicate measurements but this S value has two degrees of freedom. Thus, a triplicate measurement provides as much information about the precision of the measurement method as two dual train measurements. Similarly a quad train provides as much information as three paired trains. In curve fitting the data, weighting factors must be applied to account for the number of degrees of freedom from each data grouping.

3.3   The Relationship Between S and C

As discussed in the Appendix, a wide range of functional forms is possible for describing the relationship between standard deviation and concentration. Based on the characteristics of the available data, the form selected and justified for the ReMAP program is a simple power function as described in Equation 2,

$$S = kC^{\,p} \hspace{4cm} \text{Eq. 2}$$

where S is the estimated standard deviation for the method, C is concentration, and k and p are constants. For each measurement method, the available data are fit to this equation using a least squares regression analysis. To facilitate that regression, it is convenient to first transform equation 2 by taking logarithms to yield,

$$Ln(S) = Ln(k) + pLn(C) \hspace{4cm} \text{Eq. 3.}$$

Regression analysis yields values for $Ln(k)$ and p. The governing equation is obtained by taking inverse logs. Unfortunately, the transformation processes create yet another bias that must be accounted for.

In the database, an average value of the individual standard deviations can be calculated from the multi-train data for a selected method. The individual values of average concentration from the database can be entered into the regression equation (Eq. 3) to predict values of S at each value of C. For an unbiased model, the average value of predicted S equals the average value of S from the data. With transformation of data into the log-log plane, that criterion is not achieved. Instead the average value of Ln(S) {Predicted] will equal the average value of Ln(S) from the data. In essence, the source of this bias is that

$$\overline{Ln(S)} \neq Ln(\overline{S})$$

This bias can be accounted for by appropriately adjusting the value of k in the model. Further explanation of this bias and the procedure for bias-correction is provided in the appendix.

The above procedure provides a simple means of using available data to estimate the standard deviation as a function of average concentration for any given measurement method. At any selected concentration, the regression equation can be used to determine an estimated value of σ. Information on Est. σ and average concentration can be used to calculate relative standard deviation, the anticipated range for 99 out of 100 future individual measurements (at that stack concentration) as well as the anticipated range for the average of triplicate future measurements. It is important to note, however, that even after correcting for the various biases, the regression line is not a perfect indicator of true value of σ. This evaluation is a best estimate, based on the currently available data. Addition of new data will undoubtedly cause an adjustment to the regression equation describing the relation between Est. σ and C.

3.4   Confidence Intervals

A critical question that must be examined is "How good is the correlation?" Based on analysis of the available data, it is possible to estimate the potential bounds for the regression line through the S versus C data. These potential bounds are referred to as confidence intervals. Procedures for calculating confidence intervals are presented in the Appendix. The Appendix also includes several example calculations. Figure 4, taken directly from the Appendix, illustrates a hypothetical set of small sample, bias-corrected S versus C data. The heavy solid line through the data represents the regression line while the arched lines above and below the regression line illustrate the upper and

18

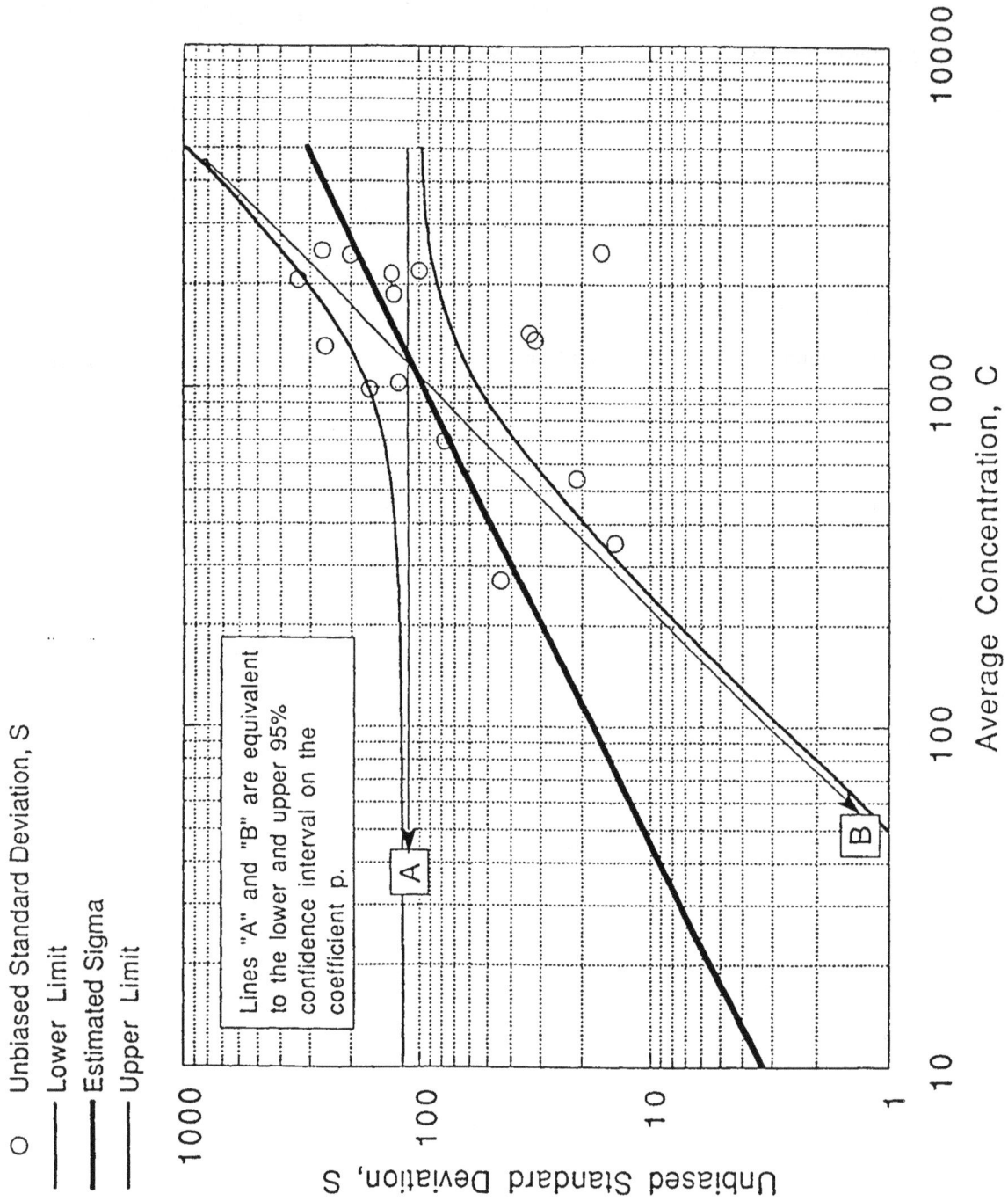

Figure 4. Regression Line and Confidence Interval for Simulated Data

19

lower confidence intervals. The regression line represents the best estimate of the relationship between σ and average concentration. The analysis used to determine the regression line also provides information on the potential bounds (at a given confidence level) of regression line. For example, the analysis provides a best estimate for the slope of the regression equation (p). The analysis also provides information to determine, at a given confidence level, the minimum and maximum slope of the relation (p). In Figure 4, the lines labeled A and B illustrate the S vs. C relation where p is at the minimum and maximum values respectively. Specifically, at the 95% confidence level, we know that the slope of the regression line is equal or greater than illustrated by Line A and less than illustrated by Line B. Similarly, the regression analysis provides information on the potential range of the leading coefficient, k. The upper and lower confidence intervals illustrated in Figure 4 represent the combined bounds on both k and p at a selected confidence level. Confidence intervals on a linear relation will always have the characteristic horn shape because of the potential range for the slope term. The true relationship between σ and true concentration cannot be determined, but it is possible to provide a best estimate of the relation. A selected level of confidence (e.g., 95 %), the regression line should not fall outside the bounds of the confidence interval.

The primary task for the Phase I ReMAP analysis is to select the appropriate value of standard deviation at given values of concentration and to use that information to determine various measures of method precision. If 95% confidence intervals are constructed around the regression line, there is 97.5% confidence that the method's standard deviation is no larger than the upper confidence bound. Similarly, there is 97.5% confidence that the standard deviation is greater than the lower confidence bound. However, neither of these confidence limits represents the best estimate on method precision. The currently available data suggests that the regression line itself is the best estimate of method precision.

Plots such as that provided in Figure 4, frequently cause difficulty for readers. Generally, a significant portion of the individual data points (circles on Figure 4) fall outside the confidence limits. This is an expected trend since the confidence intervals represent upper and lower bounds for the regression line – not upper and lower bounds for the data.

A word of caution is in order. When the statistical analysis does not require weighting, it is relatively simple to calculate confidence intervals using software routines contained in standard, commercial spreadsheet computer programs. Recall that weighting of the data is required when the individual data points have different degrees of freedom. For example, standard software can easily be used to calculate confidence intervals in situations when all of the data consist of paired train measurements. When data weighting is required, calculation of confidence intervals becomes much more complex, requiring inversion of rather messy matrices. Advanced statistical analysis computer software generally includes routines for such analyses. Alternately, special computer software will need to be written. For the current report, detailed explanation of confidence interval calculation has been limited to those situations where calculations can be performed using software routines in standard spreadsheet computer programs such as Excel.

One additional subtle issue related to calculation of confidence intervals must be addressed before proceeding with the ReMAP analysis. The appropriate calculation process depends upon how the confidence intervals are to be used. For the ReMAP study, the intended use of the various analyses is to determine Est. σ at discreet values of average concentration and to use Est. σ to calculate various precision metrics at those average concentrations. Method precision metrics are also calculated assuming that the true value of σ is at the upper and lower confidence intervals (at selected concentrations). There are alternate ways of using confidence intervals that require slightly different analysis procedures. Those procedures were carefully considered in establishing the ReMAP statistical methodology and were deemed inappropriate for the current analysis purposes.

3.5   Summary of ReMAP Analysis Process

In summary, the ReMAP analysis procedure begins with a database of available multi-train data from application of an EPA measurement method. These data are screened for outliers using procedures that purposefully try to include as much data as possible. Data should be discarded only if there is an identified problem with a measurement or if a data pair (or a single measurement from a triplicate or quad test) is demonstrably dissimilar from the remainder of the data in the data set. For each test run, the 2, 3, or 4 simultaneous measurements are entered into equation 1 to determine the standard deviation for the run. Each of these standard deviation estimates is then multiplied by the appropriate correction factor to account for the small sample bias. The array of bias corrected standard deviation data and average concentration data are weighted for the number of degrees of freedom, transformed to the Log-Log plane, and subjected to a linear regression analysis. This analysis determines values of k and p in the power function curve – equation 2. This equation is then used to determine a predicted value of standard deviation at each value of concentration. The average value of S from the test data is compared to the average value of S from the prediction to determine an appropriate value for the second bias correction factor. That factor is multiplied by the k parameter to provide an unbiased equation relating our best estimate of standard deviation to concentration. Next, the 95% confidence intervals are calculated over the range of available data. [Note, the confidence intervals are actually calculated in the Log-Log plane. The second bias correction factor is applied to the interval when it is transformed back to the Est. σ-C plane.]

Data are presented in four ways. First, the data are presented in tabular and graphical form showing a scatter plot of calculated standard deviation and relative standard deviation versus the average concentration. The second form of data presentation is a graph of the three precision metrics plotted against concentration. Each of these metrics is normalized by the concentration and is based on the best estimate of the method standard deviation. Third, data are presented to illustrate the fact that the true value of method standard deviation could be greater than or less than the best estimate. There are six curves of interest. The first two curves, representing a worst case scenario, focus on the situation that would occur if the true method standard deviation (σ) were best represented by the upper bound of the 95% confidence interval. Using that upper limit of Est.σ, the upper and lower bounds of measured concentration are calculated that encompasses 99% of future measurements. These curves, plotted against stack concentration, are denoted by the symbols C99u/S95+ and C99l/S95+. Next, the data bands encompassing 99% of future measurements are calculated, assuming that true standard deviation varies according to the regression equation. These two lines are given the notation C99u/Sbest and C99l/Sbest. Finally, consideration is given to the case where the regression analysis has provided an over estimate of standard deviation. These curves are similar to the first two but are based on the lower 95% confidence interval. These two lines, given the symbols C99u/S95- and C99l/S95-, can be considered best case scenarios. There is 97.5% confidence that the method's precision is worse than these last two lines.

For each of the plots describing measurement method precision, care has been taken to limit the range of the presentation to the range of the currently available data. There has been no extrapolation beyond the range for which experimental data was available.

The final data presentation is a table quantifying the anticipated range of future measurements at selected values of average stack concentration. These tables list C99u/Sbest and C99l/Sbest over a range of concentrations imposed or under consideration for current environmental regulations. As regards these tables as well as all other methods of describing method precision, it is important to reiterate that various parameters are not corrected to a constant excess air level.

The sections that follow examine each of the EPA Measurement Methods of interest. The first method discussed is Method 5 for determining particulate matter concentration. The precision assessment for this method is presented in great detail in hope that the reader can better understand the full scope of the assessment.

This Page Intentionally Left Blank.

4.0 EPA Particulate Matter Methods – Methods 5 and 5i

Sampling hardware used for the majority of the EPA manual isokinetic measurement methods is based upon the hardware used for measuring particulate matter (PM) concentration in stacks. The procedure for measuring stack particulate concentration has been designated EPA Method 5 and the associated hardware is referred to as a Method 5 train. (EPA, 1987) Additional details on Method 5 (and other Methods discussed in this report) can be found in 40CFR Part 60 – Appendix A under the heading for the Method. Figure 5 illustrates the Method5 hardware. Describing the key features of a Method 5 train serves as a convenient basis for further discussion of other measurement methods addressed in this study.

In general terms, application of Method 5 involves inserting a probe into a stack and extracting a composite sample that is representative of average conditions across the stack. In a typical sampling run, sample gases are extracted from the stack for a period of approximately one-hour. The stack cross section is divided into equal area segments. The probe is traversed across the stack, extracting stack flow from each segment for equal time periods. With a round stack, traversing typically occurs through two ports, located perpendicular to each other. The rate of sample extraction is adjusted such that the velocity of gases entering the probe tip is essentially equal to the local velocity of flue gas in the stack. This is referred to as isokinetic sampling. This feature of manual method sampling is included to minimize the potential for sample bias associated with preferential capture of solid phase material according to particle size.

The extracted sample is passed through a heated line to a heated filter assembly that captures solid phase particles. The mass of particulate captured during the entire sampling period is determined gravimetrically. Additional features of the method include procedures for determining the volume of flue gas extracted from the stack during the sampling period. Using standardized protocols, the particulate concentration is determined as the ratio of the mass of particulate collected divided by the volume of flue gas extracted.

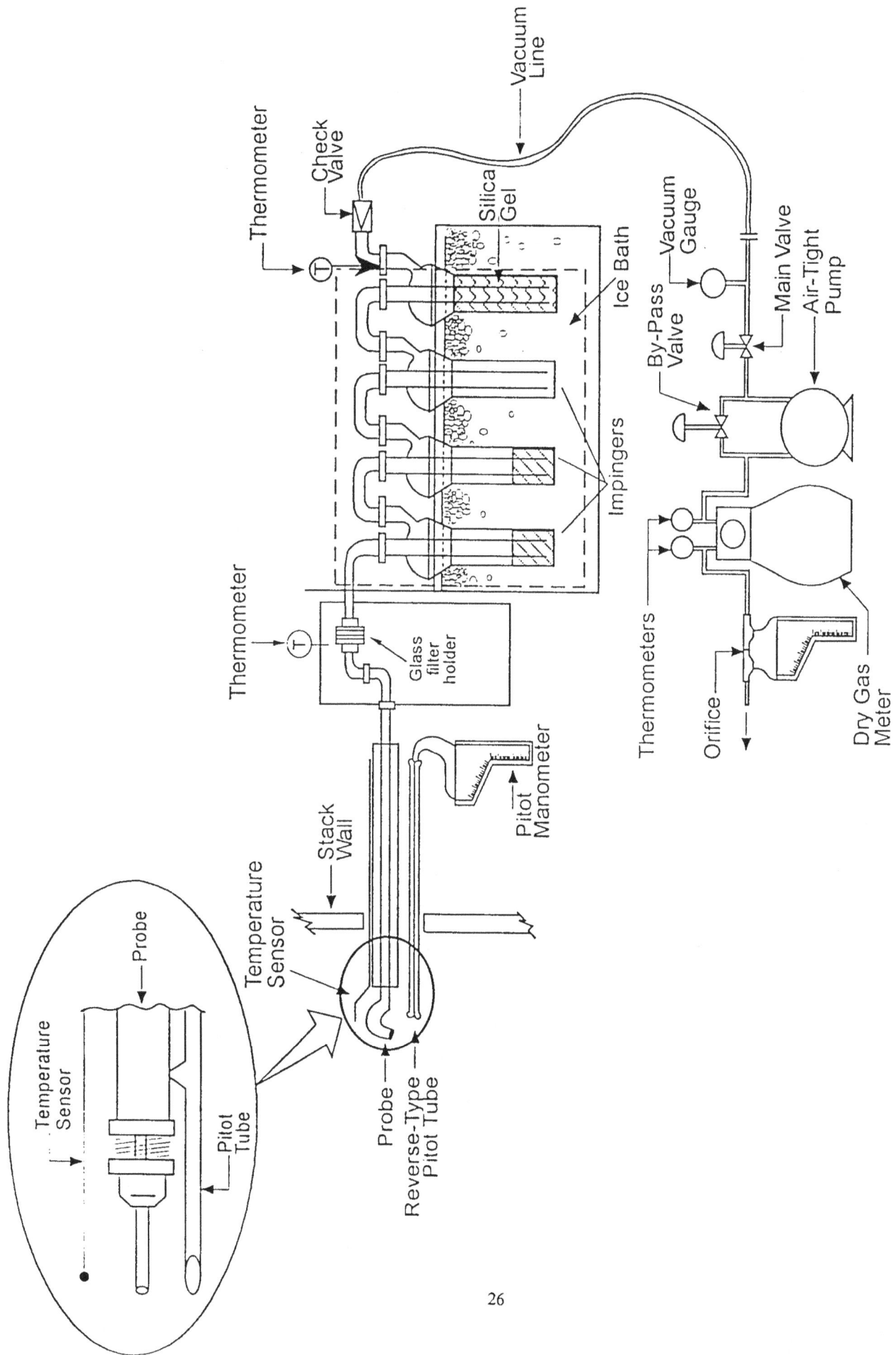

Figure 5. Schematic of Method 5 sampling train.

Figure 5 illustrates the hardware components and their arrangement for Method 5 sampling. As indicated, sample gas is extracted through a nozzle and transported in a heated glass probe to a heated filter assembly. The probe assembly consists of a glass nozzle, a heated glass probe liner, a s-type pitot probe and a thermocouple (T/C). The T/C and pitot allow determination of the local stack gas temperature and velocity, which provides a basis for adjusting the sample extraction rate to isokinetic conditions. Particulate matter in the sample may be deposited on the nozzle and probe liner walls but the majority of the particulate matter (typically >90%) is collected on a heated filter. Heating of both the probe and filter assembly is required to prevent condensation of water and other condensable materials in this portion of the sampling train (often referred to as the front half of the train). Located downstream of the heated filter box is a series of impingers in an ice bath that remove moisture from the sample. An umbilical cord connects the impingers to the meter box. The meter box contains a dry gas meter to determine the volume of dry sample extracted, means for determining pitot probe ΔP, read-outs for key temperatures, and a vacuum pump for adjusting sample extraction rate. The sampling rate is usually held between 0.5 and 1.0 cubic feet per minute.

After completion of a test run, the field technician thoroughly rinses the train components upstream of the filter with the appropriate solvent (generally acetone for particulate samples) to recover any particulate that may have been deposited on the probe walls or nozzle tip. The technician must also record a number of sampling system parameters necessary to determine the volume of gas collected and the moisture content of the flue gas. Typically an Orsat analysis is performed on the flue gas to determine the major constituents of the flue gas, particularly the oxygen concentration. Back in the laboratory, the probe rinse and filter are dried and the mass of particulate collected is determined gravimetrically. To assure that the final weight gain on the filter represents dried particulate, repeated measurements are performed. The sample is considered dry and results are reported when subsequent weighings agree within 0.5 mg. Particulate concentration is determined as the ratio of the particulate mass collected divided by the volume of flue gas collected. Usually, the sample volume is determined and reported on a dry basis and adjusted to standard temperature and pressure conditions. Standard temperature and pressure conditions used by the U S EPA are 20°C and 760 mm Hg. For regulatory purposes dilution effects are accounted for by correcting the measured concentration to a fixed percent oxygen (or carbon dioxide).

Some states require special analysis procedures to assess the mass of material that condenses in the impinger portion of the Method 5 sampling train. Those states often require that the mass of condensed phase material be combined with the particulate catch in the probe and filter to yield a total particulate phase catch. These procedures were not used for the ReMAP study. All particulate concentration data presented and analyzed in the following sections represent solid phase material collected in the front half of the train only. Moreover, conclusions drawn in this report relative to the performance of particulate measurement methods should not be applied to measurements where the analysis includes back half catch from Method 5 trains.

4.1   Method 5 Data and Precision Analysis

Multi-train data included in the ReMAP database come from three main reports. The first data set includes a series of EPA-sponsored studies conducted by Southwest Research Institute in the early 1970s to validate the particulate method (Hamil and Camann, 1974a and 1974b). Sources tested included a coal-fired power plant and two municipal waste combustors (MWCs). Tests on all three units were performed using four sampling trains operated simultaneously (quad-trains). For this study, each train was operated by a different sampling organization.

At the power plant site, testing was completed under four different operating conditions thus providing a total of 16 data points. Stack PM concentration results range from 141 to 240 mg/dscm. For the first MWC test, six test conditions were sampled providing 24 individual data points with concentrations ranging from 49 to 64 mg/dscm. The second MWC test included five quad-train runs, providing 20 individual data points. Data ranged from 103 to 161 mg/dscm. Results from these three test series are provided in Table 1. It should be noted that run numbers listed in Table 1 (and all subsequent tables listing experimental data) denote run numbers listed in the original reference. Skips in run numbers typically reflect tests where the original authors noted experimental difficulties and suggest that the run results may be invalid.

Table 1. EPA Method 5 Data - Hamil and Camann, 1974 and 1974b.

	Run Number	A	B	C	D
Power Plant	1	205	202	204	221
	2	190	196	222	185
	3	207	240	222	199
	4	155	150	188	141
MWC1	5	60.4	61.9	63.2	64.6
	6	57.1	62.2	64.1	62.5
	7	62.5	61.3	58.2	56.9
	8	54.2	51.8	49.6	45.3
	9	51	50.5	50.5	46.1
	10	56.3	59.9	62.4	51.9
MWC2	11	140	143	131	133
	12	113	107	125	128
	14	126	123	134	144
	15	153	141	161	139
	16	103	106	104	103

All data expressed as mg/dscm. Data are not corrected for oxygen content

29

The next set of multi-train Method 5 data is provided by a second EPA-sponsored study at an MWC in Dade County, Florida. The tests were directed by Southwest Research Institute (Hamil and Thomas, 1976). These tests are unique among all data collected for validation of EPA measurement methods. The stack test location provided four sampling ports located 90° apart. Dual sampling trains were used in each port providing a total of eight simultaneous measurements for each test condition. Moreover, a total of nine different sampling teams were used in the collaborative study. The experiments covered a 3-week period with the test plan calling for five runs per week. Seven different laboratories analyzed the four paired sampling trains at the end of each week. One paired train was operated by a single technician maintaining both meter boxes. The laboratory changed each week (accounting for three of the nine participating laboratories). For the remaining three paired trains, a separate laboratory operated each train for the entire 3-week span.

The test plan called for fifteen sampling runs, five per week for three weeks. Thirteen runs were actually completed; three the first week and five each the second and third week. Table 2 provides a summary of the measurement results. A total of 104 data points were collected. For run 10, a probe liner was broken on one of the eight trains (Train A operated by Laboratory 103). Accordingly, that data point was eliminated from the data analysis.

The final data source was an ASME-sponsored study by Rigo and Chandler who performed extensive multi-train experiments on a municipal waste combustor in Pittsfield, Mass (Rigo and Chandler, 1997). A total of 16 Method 5 data pairs are reported. The data range for these tests is from 14 to 74 mg/dscm, which significantly extends the overall range of the full data set. Data gathered at Pittsfield used essentially every EPA method of interest to the ReMAP program. Particulate concentration results from the Rigo and Chandler tests are provided in Table 3.

The first step in the ReMAP analysis is to determine if any of the data groups are outliers. The approach is outlined in the Appendix and includes the Dixon's-r test, a procedure taken from Statistical Process Control (SPC) methods. The Dixon's-r test is applied to individual tests consisting of three or more simultaneous measurements and is used to identify potential outliers

Table 2. EPA Method 5 Data - Dade County MWC, Hamil and Thomas, 1976.

Run Number	Lab 101		Lab 102		Lab 103		104	105	106	107	108	109
	A	B	A	B								
1	117.9	132.6					189.5	122.3	139	137.8	135.2	126.2
2	100.8	106.1					129.9	79.4	100.1	87.8	123	103.6
3	128.5	135.7					143.9	135	124.1	116.7	101.3	107.1
4			144.1	148.1			155.8	151	142.6	135.6	141.6	123.6
5			103.4	112.3			120.3	124.7	110.7	107.2	119.1	123.6
6			150.9	133.3			149.3	123.3	145.7	142.1	176.6	159.9
7			81.9	81.7			101.2	129.4	91.2	86.2	99.3	82.2
8			101.3	104.2			137.2	114.1	119.9	116.9	140.2	95.9
9					101.2	107.6	97.4	108.1	97.6	89.2	94.7	90.2
10					B.P.	158.1	158.1	144.2	146.4	140.1	136.2	171.8
11					169.9	161.3	162.8	124.4	157.6	155.1	177.4	152.8
12					254.5	190.2	164.1	146.7	152	148.1	154.7	137.8
13					188.6	189.9	185.1	157.4	159.5	154.7	163.8	139.1

All data expressed as mg/dscm. Data are not corrected for Oxygen content
B.P. denotes broken probe

Table 3. EPA Method 5 Data - Pittsfield MWC, Rigo and
Chandler 1997.

Run Number	Train A mg/dscm	Train B mg/dscm
1	40.4	43.4
3	21.3	22.2
6	14.1	14.3
7	42.9	49.7
8	28.5	25.7
9	15.9	17.4
10	29.6	27.3
11	43.2	42.1
12	27.3	20.3
13	33.9	33.1
14	29.9	28.7
15	42.8	51.6
16	40.9	44.7
17	60.7	74.3
18	43.0	42.0
19	34.5	39.4

All data expressed as mg/dscm. Data are not corrected for Oxygen content

within the test. The SPC procedure begins by breaking the data into groups representing ranges of similar concentration. The span of data is calculated for each simultaneous measurement and then weighted according to a factor that is a function of the number of simultaneous determinations (i.e., pairs, quads, etc.). Next, the average weighted span is calculated for each concentration group. If the span for a given run exceeds the weighted average span, then data from that run are abnormally large, relative to other data in that concentration range. In SPC terminology, the weighting factors are referred as D_4. Table 4 provides a listing of D_4 parameters as a function of the number of measurements in a run.

Table 4: SPC Factors for Identification of Data Outliers

Sample Size, n	D_4
2	3.267
3	2.575
4	2.282
5	2.115
6	2.004
7	1.924
8	1.864

The choice of concentration ranges is somewhat arbitrary. For the ReMAP analysis, several ranges were tested. There is a strong preference for minimizing the number of data points eliminated. Table 5 combines the data presented earlier in Tables 1, 2 and 3 and assesses the data according SPC procedures outlined above. As a first cut, the data was separated into two range groups with the first group representing data collected at the power plant, at MWC2 and at the Dade County MWC. All of these data had average PM concentrations above 94 mg/dscm. The remaining data from MWC1 and the Pittsfield MWC had average PM concentrations less than 68 mg/dscm. The weighted average spread for simultaneous measurements in the low concentration range was 13.15 mg/dscm. For measurements in the high concentration range, the weighted average data spread was 77.32 mg/dscm. Data points are suspect if the actual measurement spread is greater than these values. Data point number 17 in the Rigo and Chandler set marginally exceeds this limit. However, this data point has the highest concentration in the "low concentration" data group. When the spread on this data point is normalized by the mean concentration, the spread is on the same order as several other data points in the low group. The ReMAP program has a bias for retaining all data unless it is

Table 5. Consolidated Method 5 Data Set

Plant	Run Number	A	B	C	D	E	F	G	H	Average Concentration	Range	D4 Weighted Span
Power Plant	1	205	202	204	221					208.0	19.0	43.4
	2	190	196	222	185					198.3	37.0	84.4
	3	207	240	222	199					217.0	41.0	93.6
	4	155	150	188	141					158.5	47.0	107.3
MWC1	5	60.4	61.9	63.2	64.6					62.5	4.2	9.6
	6	57.1	62.2	64.1	62.5					61.5	7.0	16.0
	7	62.5	61.3	58.2	56.9					59.7	5.6	12.8
	8	54.2	51.8	49.6	45.3					50.2	8.9	20.3
	9	51	50.5	50.5	46.1					49.5	4.9	11.2
	10	56.3	59.9	62.4	51.9					57.6	10.5	24.0
MWC2	11	140	143	131	133					136.8	12.0	27.4
	12	113	107	125	128					118.3	21.0	47.9
	14	126	123	134	144					131.8	21.0	47.9
	15	153	141	161	139					148.5	22.0	50.2
	16	103	106	104	103					104.0	3.0	6.8
Dade County	1	117.9	132.6	189.5	122.3	139	137.8	135.2	126.2	137.6	71.6	133.5
	2	100.8	106.1	129.9	79.4	100.1	87.8	123	103.6	103.8	50.5	94.1
	3	128.5	135.7	143.9	135	124.1	116.7	101.3	107.1	124.0	42.6	79.4
	4	144.1	148.1	155.8	151	142.6	135.6	141.6	123.6	142.8	32.2	60.0
	5	103.4	112.3	120.3	124.7	110.7	107.2	119.1	123.6	115.2	21.3	39.7
	6	150.9	133.3	149.3	123.3	145.7	142.1	176.6	159.9	147.6	53.3	99.4
	7	81.9	81.7	101.2	129.4	91.2	86.2	99.3	82.2	94.1	47.7	88.9
	8	101.3	104.2	137.2	114.1	119.9	116.9	140.2	95.9	116.2	44.3	82.6
	9	101.2	107.6	97.4	108.1	97.6	89.2	94.7	90.2	98.3	18.9	35.2
	10		158.1	158.1	144.2	146.4	140.1	136.2	171.8	150.7	35.6	68.5
	11	169.9	161.3	162.8	124.4	157.6	155.1	177.4	152.8	157.7	53.0	98.8
	12	254.5	190.2	164.1	146.7	152	148.1	154.7	137.8	168.5	116.7	217.5
	13	188.6	189.9	185.1	157.4	159.5	154.7	163.8	139.1	167.3	50.8	94.7
Rigo & Chandler	1	40.4	43.4							41.9	3.0	9.9
	3	21.3	22.2							21.7	1.0	3.1
	6	14.1	14.3							14.2	0.2	0.7
	7	42.9	49.7							46.3	6.8	22.3
	8	28.5	25.7							27.1	2.9	9.3
	9	15.9	17.4							16.6	1.5	4.9
	10	29.6	27.3							28.4	2.3	7.5
	11	43.2	42.1							42.7	1.1	3.8
	12	27.3	20.3							23.8	7.0	22.8
	13	33.9	33.1							33.5	0.8	2.5
	14	29.9	28.7							29.3	1.2	3.9
	15	42.8	51.6							47.2	8.8	28.7
	16	40.9	44.7							42.8	3.8	12.4
	17	60.7	74.3							67.5	13.6	44.5
	18	43.0	42.0							42.5	1.0	3.2
	19	34.5	39.4							37.0	4.9	16.1

All data expressed as mg/dscm. Data are not corrected for Oxygen content

an obvious and significant outlier. All data points in the low concentration grouping have been retained for subsequent analysis.

The high concentration range group includes quad data and the Dade County tests using octets. Data in this range are suspect if the difference is greater than 77.32 mg/dscm. Only one measurement, run number 12 from the Dade County tests, fails to meet this criteria. Data from run number 1 in the Dade County tests is also quite large. The data report provides no indication of measurement problems associated with either of these measurements but it is obvious from inspection that, for run number 12 the two data points collected by Laboratory 103 (labeled A and B in Table 5) are higher than the other six determinations. For run number 1, data from Run C also appears abnormally high. Dixon's-r procedure was applied to the data from both runs and results indicate that test point C from run number 1 and test point A from run 12 are abnormally high and should be considered as outliers. All other data points in this data set pass the Dixon's-r criteria. After eliminating the two data points, the remaining data in the high concentration data group pass the SPA criteria.

The next step in the analysis is to calculate the average concentration and standard deviation for each group of simultaneous measurements and to correct the calculated sample standard deviations for the small sample bias. As noted in earlier discussion, the calculated value of standard deviation from the data, S, is a biased estimate of the true standard deviation, σ. Table 6 presents the correction factors used to calculate an unbiased estimate of σ as a function of the number of data points used to calculate S.

Table 6. Factors for Calculating Unbiased Estimates of σ Based on S

Sample size, n	Small Sample Correction Factor
2	1.253
3	1.128
4	1.085
5	1.064
6	1.051
7	1.042
8	1.036

Results of these calculations are presented in Table 7. Figures 6a and 6b present scatter plots of the data from Table 7. Figure 6a shows the scatter of bias corrected standard deviation versus the average particulate concentration. Figure 6b presents the same data, but in a slightly different

Table 7. Method 5 - Small Sample Bias Correction to Standard Deviation

Facility	Run No	N	Avg. Concentration	Standard Deviation	Bias Factor	Estimated Sigma
Power Plant	1	4	208.0	8.756	1.085	9.500
	2	4	198.3	16.460	1.085	17.859
	3	4	217.0	18.055	1.085	19.590
	4	4	158.5	20.502	1.085	22.245
MWC1	5	4	62.5	1.795	1.085	1.948
	6	4	61.5	3.034	1.085	3.291
	7	4	59.7	2.613	1.085	2.835
	8	4	50.2	3.783	1.085	4.104
	9	4	49.5	2.295	1.085	2.491
	10	4	57.6	4.565	1.085	4.953
MWC2	11	4	136.8	5.679	1.085	6.162
	12	4	118.3	9.912	1.085	10.755
	14	4	131.8	9.394	1.085	10.193
	15	4	148.5	10.376	1.085	11.258
	16	4	104.0	1.414	1.085	1.534
Dade County	1	7	130.1	8.120	1.042	8.461
	2	8	103.8	16.601	1.036	17.199
	3	8	124.0	14.776	1.036	15.308
	4	8	142.8	9.914	1.036	10.271
	5	8	115.2	7.873	1.036	8.157
	6	8	147.6	16.190	1.036	16.773
	7	8	94.1	16.207	1.036	16.790
	8	8	116.2	16.097	1.036	16.676
	9	8	98.3	7.107	1.036	7.363
	10	7	150.7	12.504	1.042	13.029
	11	8	157.7	15.655	1.036	16.219
	12	7	156.2	16.994	1.042	17.707
	13	8	167.3	18.541	1.036	19.208
Rigo & Chandler	1	2	41.9	2.133	1.253	2.673
	3	2	21.7	0.677	1.253	0.849
	6	2	14.2	0.144	1.253	0.181
	7	2	46.3	4.827	1.253	6.048
	8	2	27.1	2.015	1.253	2.525
	9	2	16.6	1.063	1.253	1.332
	10	2	28.4	1.628	1.253	2.040
	11	2	42.7	0.812	1.253	1.017
	12	2	23.8	4.939	1.253	6.189
	13	2	33.5	0.548	1.253	0.686
	14	2	29.3	0.846	1.253	1.060
	15	2	47.2	6.213	1.253	7.784
	16	2	42.8	2.675	1.253	3.352
	17	2	67.5	9.636	1.253	12.074
	18	2	42.5	0.701	1.253	0.879
	19	2	37.0	3.485	1.253	4.367

All data expressed as mg/dscm. Data are not corrected for Oxygen content

Figure 6a. EPA Method 5 Data - Standard Deviation

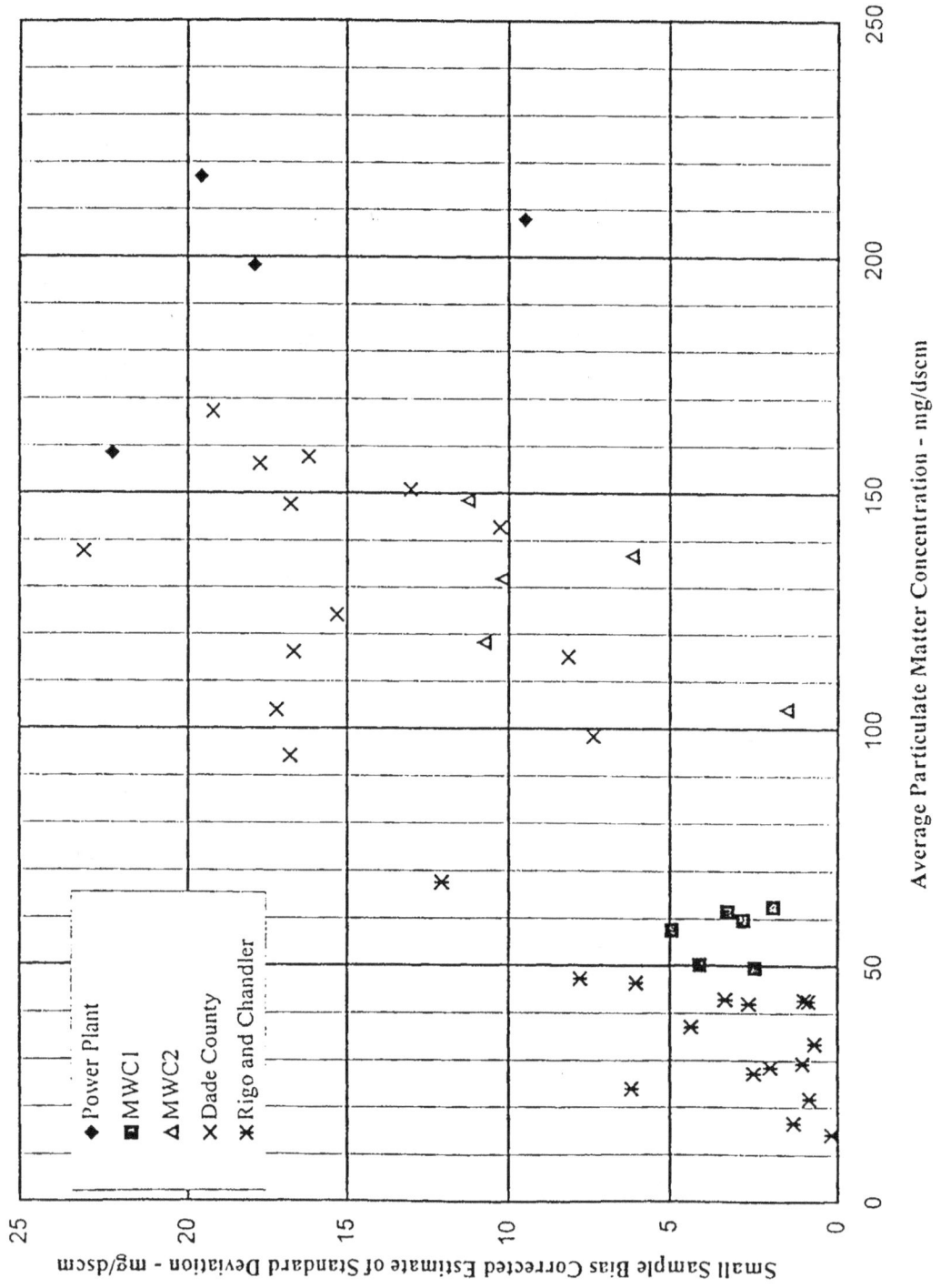

Small Sample Bias Corrected Estimate of Standard Deviation - mg/dscm

Average Particulate Matter Concentration - mg/dscm

Legend:
◆ Power Plant
■ MWC1
△ MWC2
✕ Dade County
✳ Rigo and Chandler

37

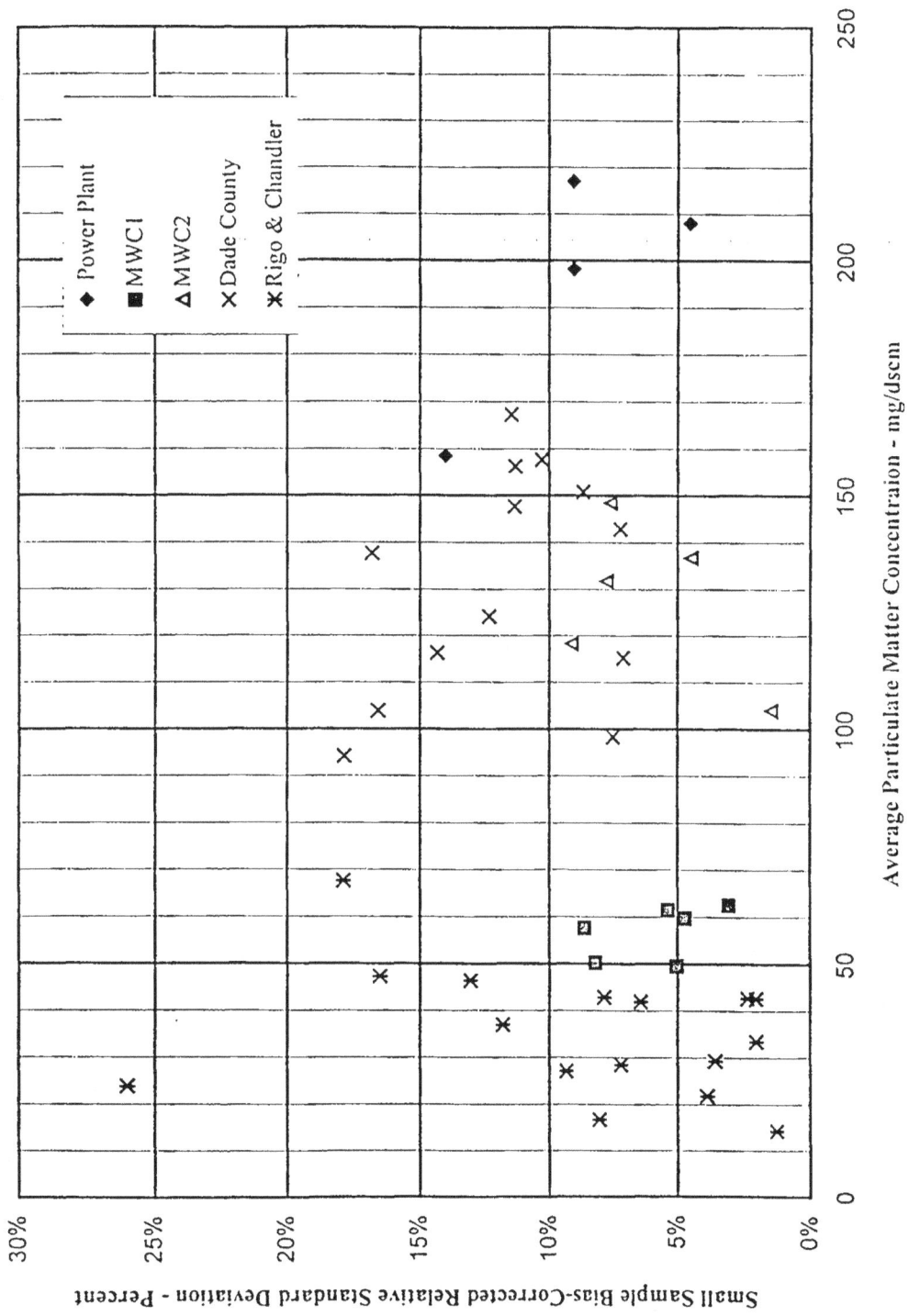

Figure 6b. EPA Method 5 Data - Relative Standard Deviation

38

format. Here the standard deviation has been normalized by the concentration and presented in units of percent.

The next portion of the data analysis is to evaluate the relationship between the estimated standard deviation and the average particulate concentration. However, before performing the regression analysis, it is necessary to weight the data according to the number of degrees of freedom for each measurement group. As discussed in the previous section, statistical assessment of data containing differing degrees of freedom involves complex matrix inversion procedures. Details of the calculation procedure are not included here.

It is instructive to review the rationale for weighting the data. With a quad-train, the four individual particulate measurements can be used in a variety of ways. For example, Train A can be grouped with Trains B, C, and D to calculate three different standard deviations or the four measurements can be combined in the calculation of a single value of S. Determination of S based on data from Trains A and B provides the same level of information achieved from dual train testing. Similarly, S determinations using data from Trains A and C or from Trains A and D convey the same level of information as dual train measurements. Clearly, a calculated value of S using all four simultaneous measurements contains more information than a calculation based on two measurements. When a data set contains results from dual, triple, quad, etc. measurements it is necessary to weight the various data to account for the relative quantity of information provided in each test.

The weighted data set is then fit to a power function relationship, as presented below.

$$S = kC^p \qquad\qquad \text{Eq. 1}$$

To assist in that regression, the data is transformed into the log-log plane such that the governing equation becomes:

$$Ln(S) = Ln(k) + pLn(C) \qquad\qquad \text{Eq. 2}$$

By performing the transformation, the regression analysis is linearized. Results from the analysis of data in Table 7 are summarized in Table 8 below.

Table 8. Method 5 Regression Analysis Results

Parameter	Analysis Result
p	1.3063
Ln(k)	-3.8985
k	0.02027
Standard error for p	0.1477
Degrees of freedom	42
Sum of Residuals	5.173 E-14
Weighted Sum of Squares of Residuals	39.878
t	8.84
Standard Deviation of Residuals	0.9744

A regression analysis is a mathematical procedure that yields a best estimate for the curve fit parameters. One critical question is whether the indicated values of k and p are statistically significant. One approach to answering this question is the Student-T test. Standard statistical tables list the t-statistic as a function of the confidence level (e.g. 95% confidence) and the number of degrees of freedom. The regression analysis also produces a value of the t parameter. If the calculated t-parameter is greater than the critical t-statistic from the tables, then the regression results are statistically significant. Conversely, if the calculated t-parameter is less than the critical t-statistic, then the regression analysis results could have occurred by random chance. As shown in Table 8, the calculated value of the t parameter is 8.84, which is well above 2.020, the critical-t parameter for 42 degrees of freedom, at the 95% confidence level. The large relative value of t assures that there is a relationship between S and C and that the relationship, as indicated by the regression analysis, did not occur by chance.

Transformation of data from the real plane to the log-log plane greatly eases the regression analysis but it introduces a potentially significant bias to the results. One characteristic of a linear regression analysis is that the average of the predicted values for the dependent variable should equal the average value from the actual data. Since the regression was performed in the log-log plane, the weighted average value of Ln(S) will be the same for both the actual data and the predictions. However, the average value of the predicted values of S will not necessarily be equal to the average of the small sample bias corrected S values. For the current data set, the sum of the individual, small sample bias corrected S values is 364.93 while the sum of the predicted values of standard deviation (at the observed values of concentration) is 351.013. Thus, the predicted values of S are biased low and a correction factor must be applied. For the Method 5 data, the log transformation correction factor is 364.93/351.013 = 1.0397.

The equation describing the estimated values of standard deviation versus concentration is the best estimate available, based on available multi-train experimental data, but there is uncertainty associated with this equation. The slope of the regression line (p) and the value of the leading constant (k) may be greater or smaller than predicted. Statistical data in Table 8 can be used to quantify uncertainty in the regression equation. Specifically, the 95% confidence intervals on the regression equation will be calculated.

The 95% confidence interval on the slope term can be expressed as

$$P_{95\%} = P_{predicted} \pm t_{95\%} * [SE(coeff)] \hspace{3cm} Eq.\ 3$$

where $P_{95\%}$ represents the upper and lower bounds of the slope coefficient, $t_{95\%}$ is the critical t-statistic at the 95% confidence level and the appropriate number of degrees of freedom, and SE(coeff) is the standard error of the coefficient.

As indicated in Table 8, the predicted value of the power term in the regression equation (p) is 1.3063 and the standard error of that coefficient is 0.1477. The critical t statistic for 42 degrees of freedom, at the 95% confidence level is 2.020 (available from standard statistical tables). Thus, the best estimate for the slope of the regression line is the predicted value (1.3063) but with 95% confidence it can only be concluded that the value of the p coefficient is between 1.008 and 1.605. Implications of the potential range of this slope term are discussed in more detail later.

The weighted least squares numerical analysis provided information necessary to determine confidence intervals on the regression equation. Results of those calculations (at the 95% confidence level) are presented in Figure 7. When plotted on log-log scale, the regression equation is a straight line and the confidence intervals appear as horn shaped curves on either side of the prediction. All three lines in this figure have been adjusted to include the log-log transformation

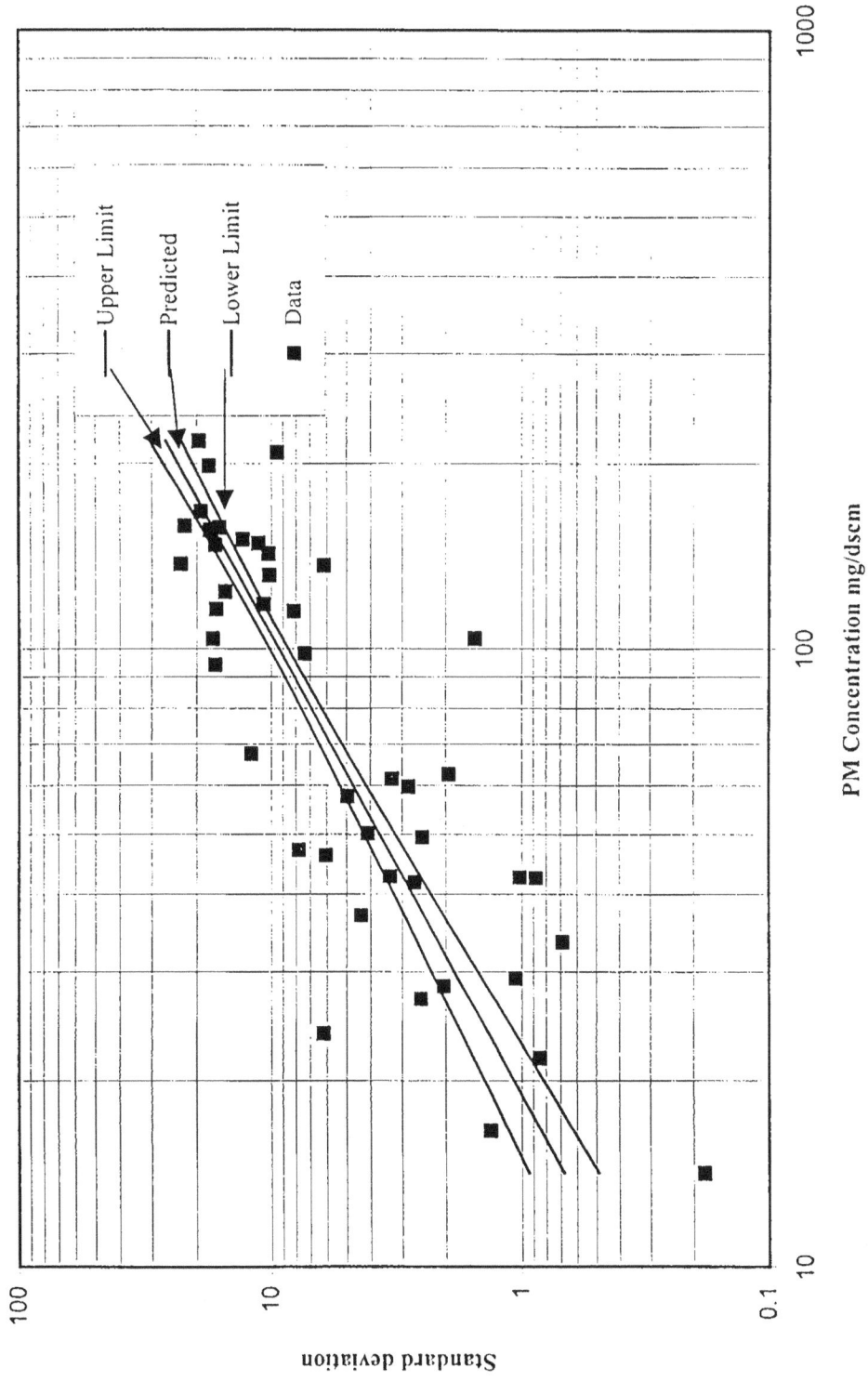

Figure 7. Regression Line and 95% Confidence Interval for EPA Method 5

bias correction factor. Superimposed on Figure 7 is the small sample bias corrected data from Table 7.

The meaning of confidence intervals often confuses those with limited background in statistical analysis. The straight line through the data represents the best estimate of the relationship between standard deviation σ and mean concentration, μ. The confidence intervals define potential bounds for the regression line – the straight line. Confidence intervals do not represent boundaries for the actual data. It is possible to calculate potential bounds for data, but those bounds are referred to as tolerance intervals. Thus, it is fully anticipated that a portion of the experimental data (individual determinations of S) will fall outside the confidence intervals.

Before proceeding with additional assessment of the Method 5 results, it is necessary to examine the implications of the regression analysis. The regression equation itself was found to be

$$S = 0.0211C^{1.306}$$ Eq. 4

If both sides of the equation are divided by the mean concentration, the left-hand term becomes S/C, which is the relative standard deviation (RSD). After performing this operation, the regression equation becomes

$$RSD = 0.0211C^{0.306} \times 100\%.$$ Eq. 5

This implies that the RSD increases with increasing concentration, which is a difficult result to rationalize. Typical random errors that might be attributed to the sample collection process, such as failure to adequately rinse particulate matter from the probe liner, should produce errors that are roughly proportional to the PM loading. Another, often observed error in sample collection is for a small portion of the filer to stick to the filter housing. This type of error causes an underestimation of the mass of particulate collected but the magnitude of the error will not be a function of concentration. Random error in the weighing process should also be relatively independent of PM concentration. It can even be argued that the relative magnitude of analytical error might decrease with increasing concentration. These considerations suggest that the value of the slope term (for any

Method) should be expected to fall between zero and 1.0. More significantly, it is suggested that the slope of the regression line in Figure 7 and in Equations 4 and 5 is too high.

The forgoing statistical analysis is obviously based on the available data and is influenced by the characteristics of those data. Two factors should be noted. First, the analysis indicated a relatively large value for the Standard Error of the p term. The 95% confidence interval on p ranged from 1.008 to 1.605. This lower limit on the slope term is tantalizingly close to 1.0, suggesting that the anticipated bounds on the regression equation (based on physical considerations) have been captured by the confidence interval of the statistical analysis.

A separate argument has been forwarded, suggesting that it may not be valid to group the various Method 5 data sets into a single analysis. The mathematical procedures of regression analysis will predict a high value for the slope term if the majority of the data at high concentration have excessively high S or if the data at lower concentrations have uncharacteristically low values of S. In the outlier analysis presented earlier, the available data were divided into two concentration groups. The high concentration group contained quad train and octet data with all tests reporting average PM concentration above 94 mg/dscm. These data were collected in the early to mid 1970s. The low concentration data included six quad train runs from MWC 1 and 16 paired train runs from the tests at Pittsfield. The Pittsfield data was collected in the mid-1990s. The regression analysis is heavily weighted by the Dade County octet data that is also high concentration data. If the Dade County data exhibited uncharacteristically high standard deviation, the slope term from the regression analysis would be uncharacteristically high. It has been suggested that the high concentration data were collected shortly after Method 5 was first developed and that the field testing crews were still learning how to properly apply the Method. In contrast, the Pittsfield data was collected more than a decade later, allowing the testing firm to refine sampling procedures.

A second consideration involves the stack sampling time for the data include in Table 5. With the exception of the Rigo and Chandler data set, sampling times were nominally set at one-hour. However, for the tests at Pittsfield, stack gases were collected for approximately 4 hours. This difference in sampling time could possibly result in lower S values for data collected at lower particulate concentration.

There is certainly merit to an argument that the skill level of sampling teams directly impact the standard deviation of measurement results. However, there is no direct information available to quantify the capability of testing teams or to provide relative weighting of data quality. Similarly, it is reasonable to speculate that sampling time might impact measurement precision but there is nothing within Method 5 that precludes extended sample collection times. For these reason, the analysis of the Method 5 data will continue based on the entirety of the available data but with a strong caution that slope of the true S versus C relation is probably very close to 1.0.

Not withstanding the forgoing comments, the regression line in Figure 7 provides the best estimate available for the standard deviation of data collected using Method 5. As discussed in Section 3, this estimate of standard deviation also defines the anticipated distribution of future measurements collected with that method. For example, based on currently available data, it is anticipated that repeated Method 5 measurements of a stack gas containing 100 mg/dscm of particulate matter would exhibit a standard deviation of 8.639 mg/dscm or 8.64 % RSD. This is a hypothetical source, where the stack PM concentration is not varying with time. Sixty eight percent of future measurements taken on this stack should fall within 1.0 σ ($\pm$8.639 mg/dscm) and ninety-nine percent of those measurements should fall within the range of $\pm$ 2.57*σ. Thus, in the example of a stack with a PM loading of 100 mg/dscm, 99 out of 100 Method 5 measurements are expected to fall within $\pm$ 22.20 mg/dscm of the true concentration. For the average of triplicate measurements, 99 out of 100 measurements would fall within the range of $\pm$ 2.57*σ/$\sqrt{3}$. Thus, the average of triplicate Method 5 measurements from this hypothetical stack is expected to fall within $\pm$ 12.82 mg/dscm of the true concentration.

Figure 8 presents the predicted relative standard deviation and the 99% bounds for future single measurements as a function of stack PM concentration. Data in this figure are based on currently available data and do not include the effect of time variation in source characteristics. The X-axis of Figure 8 represents the true concentration of PM in the hypothetical stack. Values indicated on the Y-axis represent the precision of Method 5 at the selected values of stack concentration. The 99% bounds are also normalized by the stack concentration and represent the anticipated range of individual measurements. Compliance with regulatory limits is typically based on the average of

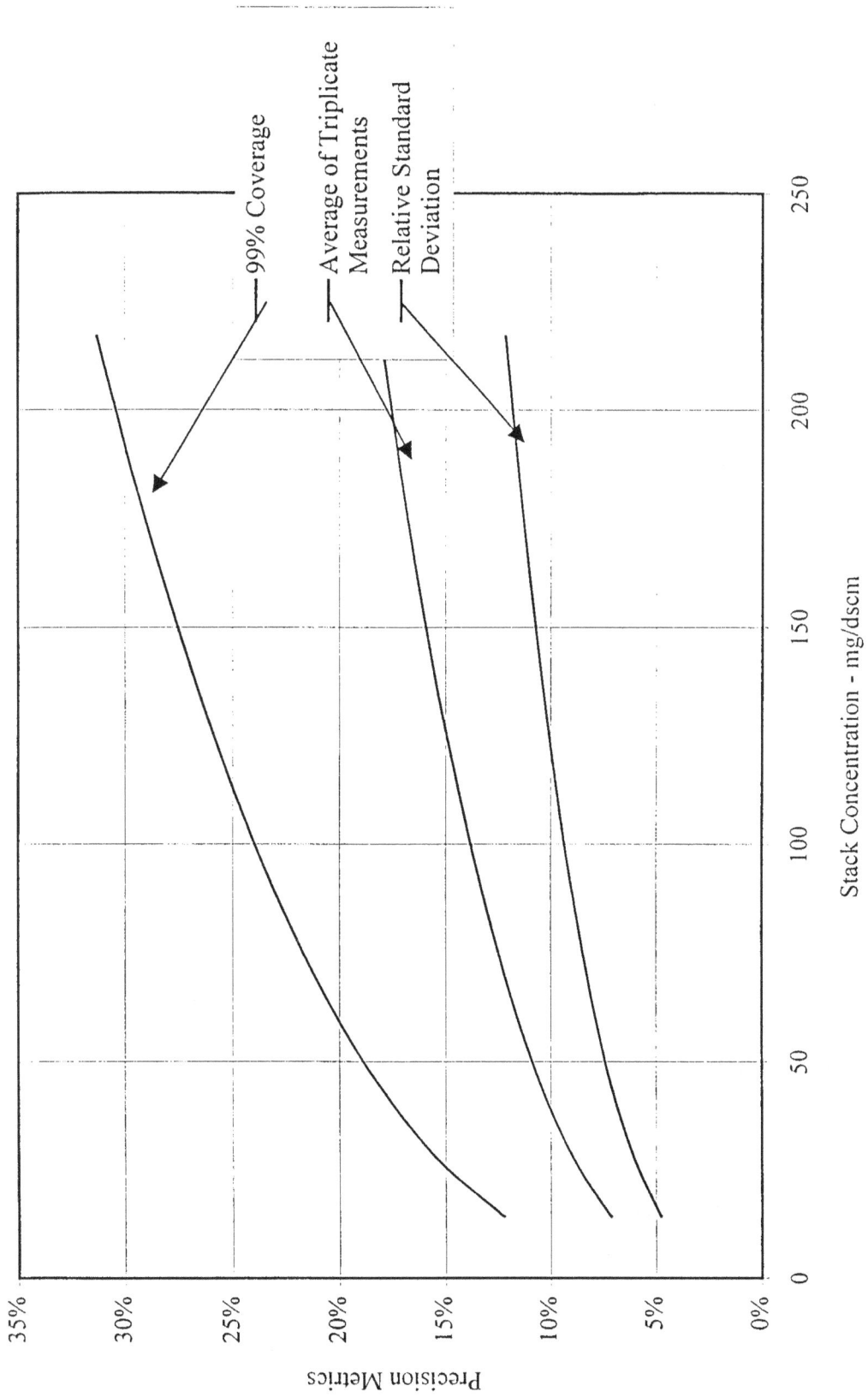

Figure 8. Weighted. EPA Method 5 precision

triplicate measurements. The predicted range for 99 out of 100 triplicate measurements is also included in Figure 8. When Method 5 is applied to a real stack, a wider range of experimental results can be anticipated due to time variations in source characteristics. Data presented in Figure 8 should not be extrapolated beyond the indicated limits. Further, based on physical considerations, it is anticipated that the true variation of these precision metrics with concentration are expected to be a flat line or even to decrease slightly (whereas the curves increase) with increasing concentration.

Data in Figure 8 were generated using the predicted values of standard deviation. Recall, however, that there is uncertainty in those estimates. It is know with 95% confidence that the relationship between standard deviation and concentration falls between the upper and lower confidence limits illustrated in Figure 7. If the actual relationship between σ versus C for Method 5 conforms to the upper confidence limits, the anticipated range of future Method 5 data will be greater than suggested by the data in Figure 8. Conversely, a tighter range of concentrations are anticipated if the variation in standard deviation conforms to the lower confidence limit. Figure 9 illustrates the ranges of anticipated concentration data under three scenarios: (1) when standard deviation conforms to the upper confidence limit; (2) when Est. σ conforms to the predicted relationship; and (3) when Est. σ conforms to the lower confidence limit. The X-axis in Figure 9 represents the true concentration of PM in a hypothetical stack that does not vary with time. The Y-axis represents the anticipated range of measured concentrations using Method 5. The upper and lower curves in the figure represent the upper and lower bounds for 99 out of 100 future measurements, assuming that the standard deviation equals the upper 95% confidence limit. There is 97.5% confidence[3] that 99 out of 100 future measurements would fall below the upper curve and 97.5% confidence that the future measurements will fall above the lower curve. Similar curves are provided for the cases where Est. σ conforms to the regression curve fit and where Est. σ is equal to the lower confidence limit.

Based on the above analysis, and concerns over the slope of the regression equation, it is difficult to draw firm conclusion about the actual precision of Method 5. However, certain trends do appear obvious. Within the confidence bounds of the analysis and based on the available data, it appears that Method 5 standard deviation varies approximately linearly with concentration and that the

---

[3] 95% confidence implies that there is a 2.5% chance that the σ relationship falls above the upper confidence limit and a 2.5% chance that the relationship falls below the lower confidence limit.

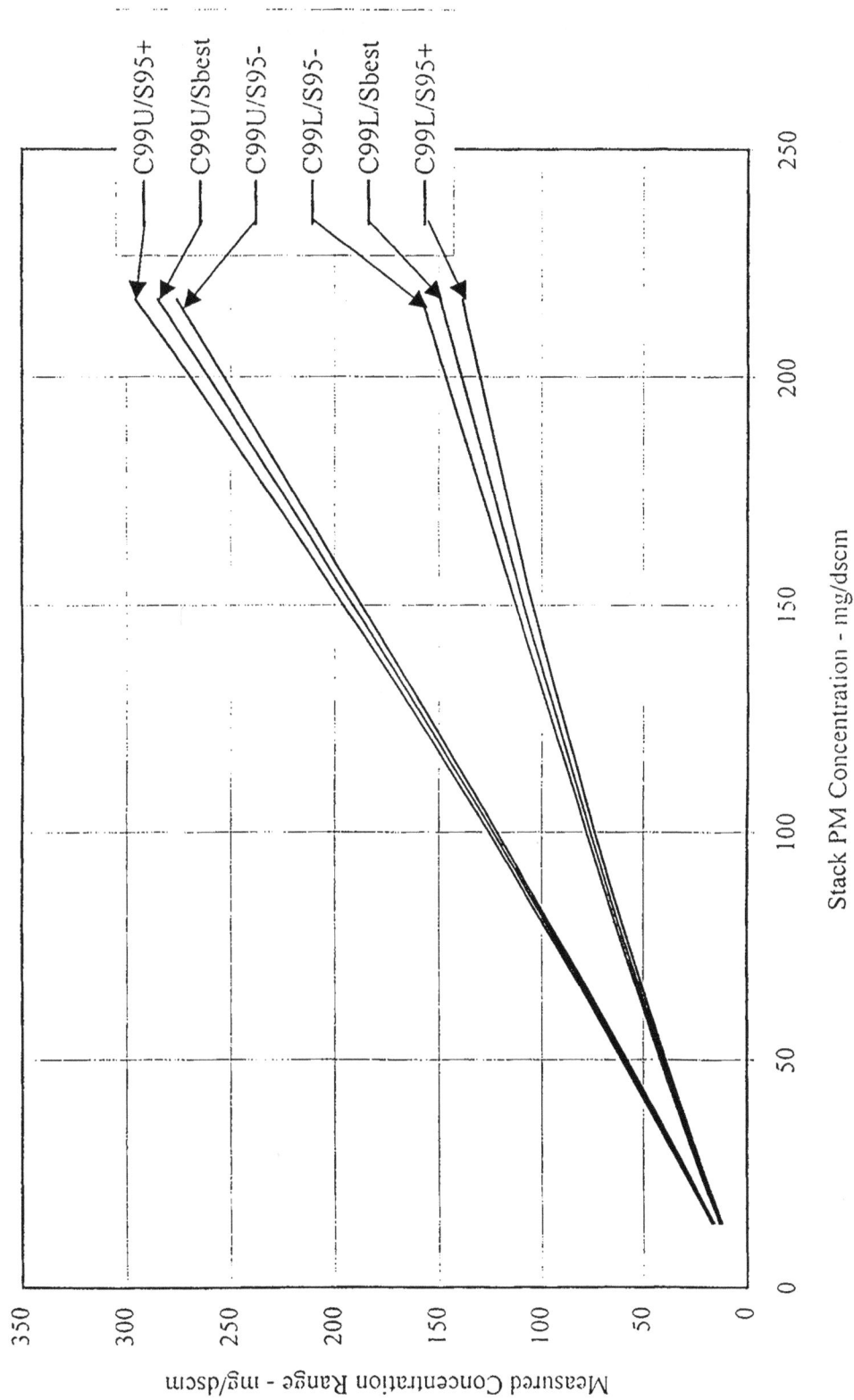

Figure 9. Estimated Data Ranges for Measurements Using EPA Method 5.

relative standard deviation for the method is approximately constant. For PM concentrations between 15 and 217 mg/dscm, the best estimate of the relative standard deviation for Method 5 is between about 4.8% to 12.2%.

4.2   Method 5i Data

In the middle 1990s the U.S. EPA began development of a new particulate measurement method, specifically designed to improve measurement precision at low loadings. The method itself was published in 1999 as part of the new MACT regulations governing hazardous waste incinerators (62 FR 52828, Sept. 30, 1999) and has been given the designation Method 5i. The hardware configuration for Method 5i is illustrated in Figure 10. Methods 5 and 5i are similar in many respects, but there are two important hardware differences and several operational differences. The primary hardware differences are in the filter assembly for the two methods. Method 5 uses a large diameter filter that must be carefully removed from its holder as part of the sample recovery process. Often a small quantity of the collected particulate can be lost or a small portion of the filter itself can adhere to the holder walls. This results in measurement imprecision that can potentially become critical when the total particulate catch is small. Method 5i uses a much smaller diameter filter and filter holder. The recovery and analysis procedures call for the filter to remain in its holder through the entire weighing process. This eliminates certain sources of random error but it creates another potential problem. Since the weight of the glass filter holder is much larger than the weight of the collected particulate, the analysis process must determine a small weight gain in a relatively large mass. Because of the small filter diameter, Method 5i is intended for use only under situations where the particulate concentration is expected to be below 50 mg/dscm. The second key feature implemented with Method 5i is the requirement that tests be conducted using dual trains. Moreover, measurement precision requirements are defined as part of the method.

Data on the precision of Method 5i comes from two studies directed primarily at evaluation of particulate matter continuous emission monitors. The first of these studies, conducted under EPA sponsorship, was executed on a hazardous waste incinerator owned by Dupont and located in Wilmington, Delaware (62 FR 67788). The second study, sponsored by an industry consortium, was conducted at a hazardous waste incinerator owned by the Eli Lilly Company (Eli Lilly, 1999). Results from these two studies provide a large database for assessment of Method 5i precision. Note that there are numerous experimental programs that were recently completed (or still underway) using Method 5i for calibration of PM continuous emission monitoring systems. Since Method 5i

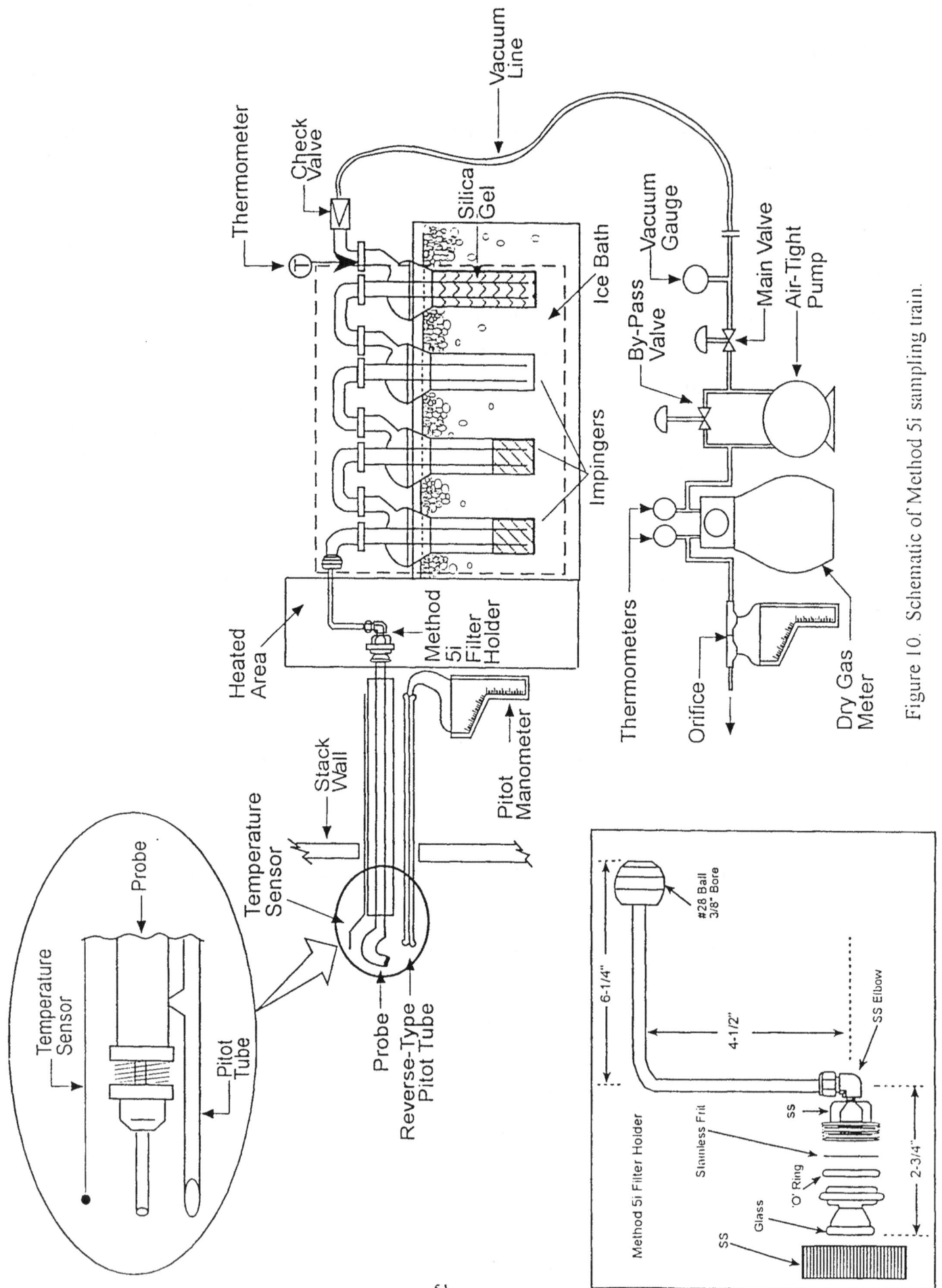

Figure 10. Schematic of Method 5i sampling train.

Vacuum Line

Thermometer

Check Valve

Silica Gel

Ice Bath

Vacuum Gauge

By-Pass Valve

Main Valve

Air-Tight Pump

Impingers

Thermometers

Orifice

Dry Gas Meter

Heated Area

Method 5i Filter Holder

Stack Wall

Temperature Sensor

Pitot Manometer

Probe

Reverse-Type Pitot Tube

Temperature Sensor

Probe

Pitot Tube

Method 5i Filter Holder

#28 Ball 3/8" Bore

6-1/4"

4-1/2"

SS Elbow

2-3/4"

Stainless Frit

ss

'O' Ring

Glass

SS

51

requires use of dual trains, the available database for assessment of this method's precision is expected to greatly expand over time.

As noted above, Method 5i is a relatively new measurement method and there are a variety of subtle issues associated with execution of the method that significantly impact the precision of results. Both the EPA test report and the Eli Lilly report note that there was a significant learning curve associated with obtaining acceptable data. In both cases a significant number of paired train tests were performed but have not been used by either group for subsequent analysis. For the ReMAP program, the only data used are results from tests considered valid by the original study authors. Tables 9a and 9b provide a summary of results from both studies. Included in the tables are the actual data pairs, the average concentration for each test run, calculated standard deviation for each pair, and the small sample bias-corrected standard deviation results.

Figures 11a and 11b present the data from Table 9a and b as scatter plots of the Method 5i standard deviation data versus average PM concentration. Both figures are based on the small sample bias-corrected standard deviation. Fig. 11a presents standard deviation in units of mg/dscm while Figure 11b presents relative standard deviation. Note that the data spans a range from less than 10 to almost 50 mg/dscm. Also note the general character of the data spread. Over the entire concentration range, the individual estimates of standard deviation are broadly distributed with no discernable trend to either standard deviation or relative standard deviation.

The data in Tables 9a and 9b were subjected to outlier analysis using SPC screening criteria. Several concentration range grouping were examined attempting to group those data at the very low particulate concentrations. In general, the screenings suggested potentially abnormal data spread for runs 53, 64, 66 and 71 from the Eli Lilly tests and run 60 from the EPA Dupont tests. Examination of these data points suggests that runs 64, 66, and 71 from the Eli Lilly tests are only marginally above the SPC screening criteria. Accordingly, those data points were retained for subsequent analysis. The spreads for data point 53 (Eli Lilly) and 60 (EPA Dupont) were sufficiently large that they have been deleted from the following analysis.

Table 9a. Method 5i Data and Standard Deviation - Eli Lilly Data

Run Number	Train A	Train B	Avg Concentration mg/dscm	Standard Deviation	RSD	S - Bias Corrected	RSD - Bias Corrected
2	43.1	43.2	43.2	0.1126	0.26%	0.1411	0.33%
4	30.7	30.3	30.5	0.2897	0.95%	0.3630	1.19%
5	32.0	29.9	30.9	1.4426	4.66%	1.8075	5.84%
6	21.1	19.2	20.2	1.3265	6.58%	1.6621	8.24%
7	26.9	21.7	24.3	3.7034	15.26%	4.6404	19.12%
8	29.1	26.1	27.6	2.0954	7.58%	2.6255	9.50%
9	28.1	29.6	28.9	1.1086	3.84%	1.3890	4.81%
10	30.7	26.4	28.5	3.0113	10.55%	3.7732	13.22%
11	31.1	31.0	31.0	0.0959	0.31%	0.1201	0.39%
12	27.0	23.2	25.1	2.6604	10.59%	3.3334	13.27%
13	28.6	26.8	27.7	1.2587	4.55%	1.5771	5.70%
14	27.3	26.1	26.7	0.8514	3.19%	1.0668	4.00%
15	28.3	28.6	28.5	0.1454	0.51%	0.1822	0.64%
16	29.4	30.7	30.1	0.9726	3.24%	1.2187	4.06%
18	31.6	30.1	30.8	1.0581	3.43%	1.3258	4.30%
19	31.4	31.5	31.5	0.1014	0.32%	0.1271	0.40%
20	31.7	30.9	31.3	0.5755	1.84%	0.7211	2.31%
21	31.2	30.9	31.1	0.2468	0.79%	0.3092	1.00%
22	35.3	40.2	37.7	3.4671	9.19%	4.3443	11.52%
23	33.0	31.9	32.5	0.7770	2.39%	0.9736	3.00%
24	29.6	26.3	28.0	2.3261	8.32%	2.9146	10.43%
25	25.9	25.8	25.9	0.0774	0.30%	0.0970	0.38%
26	25.7	25.6	25.7	0.0920	0.36%	0.1153	0.45%
27	27.8	28.8	28.3	0.6970	2.46%	0.8734	3.08%
28	28.7	27.7	28.2	0.6602	2.34%	0.8272	2.93%
29	28.6	28.9	28.7	0.2055	0.72%	0.2575	0.90%
30	28.3	28.6	28.4	0.2176	0.77%	0.2727	0.96%
31	26.3	27.2	26.8	0.6770	2.53%	0.8482	3.17%
32	26.9	25.6	26.3	0.8748	3.33%	1.0962	4.18%
33	28.8	28.5	28.7	0.1924	0.67%	0.2411	0.84%
34	29.5	28.8	29.1	0.4295	1.47%	0.5381	1.85%
35	29.3	28.8	29.1	0.3468	1.19%	0.4346	1.49%
36	29.2	28.4	28.8	0.5606	1.94%	0.7024	2.44%
37	31.3	31.5	31.4	0.1460	0.47%	0.1829	0.58%

Table 9a (Continued). Method 5i Data and Standard Deviation - Eli Lilly Data

Run Number	Train A	Train B	Avg Concentration mg/dscm	Standard Deviation	RSD	S - Bias Corrected	RSD - Bias Corrected
38	25.5	24.9	25.2	0.4379	1.74%	0.5486	2.18%
39	26.6	27.6	27.1	0.6536	2.41%	0.8189	3.02%
40	24.5	25.5	25.0	0.7449	2.98%	0.9333	3.73%
41	25.2	26.1	25.7	0.6842	2.67%	0.8572	3.34%
42	22.9	23.0	22.9	0.0539	0.23%	0.0675	0.29%
43	23.4	23.6	23.5	0.1532	0.65%	0.1920	0.82%
44	26.6	30.2	28.4	2.4926	8.77%	3.1232	10.99%
45	33.7	33.3	33.5	0.3306	0.99%	0.4143	1.24%
46	35.2	35.4	35.3	0.1247	0.35%	0.1563	0.44%
47	32.9	32.6	32.7	0.2610	0.80%	0.3271	1.00%
48	36.6	34.6	35.6	1.4279	4.01%	1.7891	5.02%
49	34.5	34.4	34.4	0.1079	0.31%	0.1352	0.39%
50	35.1	37.2	36.2	1.5060	4.16%	1.8870	5.21%
51	40.1	39.8	39.9	0.1523	0.38%	0.1908	0.48%
52	38.7	39.4	39.1	0.5131	1.31%	0.6429	1.65%
53	30.3	37.7	34.0	5.2347	15.40%	6.5591	19.30%
54	37.9	37.9	37.9	0.0135	0.04%	0.0170	0.04%
55	37.7	37.7	37.7	0.0217	0.06%	0.0272	0.07%
57	51.9	51.0	51.5	0.5948	1.16%	0.7453	1.45%
58	49.1	48.6	48.9	0.3423	0.70%	0.4289	0.88%
59	48.4	49.8	49.1	0.9959	2.03%	1.2479	2.54%
60	43.8	46.8	45.3	2.1201	4.68%	2.6565	5.86%
61	44.1	47.9	46.0	2.6992	5.86%	3.3821	7.35%
62	45.4	45.9	45.6	0.3403	0.75%	0.4263	0.93%
64	42.0	48.0	45.0	4.2466	9.44%	5.3210	11.83%
65	47.6	49.1	48.3	1.0108	2.09%	1.2665	2.62%
66	41.5	47.1	44.3	3.9828	8.99%	4.9905	11.26%
69	43.4	41.7	42.5	1.2008	2.82%	1.5046	3.54%
70	40.9	42.8	41.9	1.3152	3.14%	1.6479	3.94%
71	41.7	35.9	38.8	4.0443	10.42%	5.0675	13.06%
72	39.2	43.3	41.3	2.8800	6.98%	3.6087	8.74%
73	41.3	43.6	42.4	1.6364	3.86%	2.0504	4.83%
74	42.8	46.2	44.5	2.3755	5.34%	2.9765	6.69%

Table 9b. Method 5i Data and Standard Deviation - EPA Dupont Data

Run Number	Train A	Train B	Avg Concentration mg/dscm	Standard Deviation	RSD	S - Bias Corrected	RSD - Bias Corrected
50	5.45	6.04	5.75	0.4231	7.36%	0.5301	9.23%
51	3.44	4.74	4.09	0.9233	22.57%	1.1569	28.28%
52	10.51	9.68	10.09	0.5857	5.80%	0.7339	7.27%
53	6.24	7.05	6.65	0.5706	8.59%	0.7150	10.76%
54	13.01	12.56	12.79	0.3219	2.52%	0.4033	3.15%
55	10.79	8.72	9.76	1.4665	15.03%	1.8375	18.83%
56	10.81	13.54	12.17	1.9277	15.83%	2.4154	19.84%
57	11.11	12.95	12.03	1.3028	10.83%	1.6324	13.57%
58	13.55	11.91	12.73	1.1563	9.08%	1.4488	11.38%
59	12.02	11.77	11.90	0.1779	1.50%	0.2229	1.87%
60	11.20	6.27	8.73	3.4892	39.95%	4.3720	50.06%
62	16.18	16.81	16.50	0.4419	2.68%	0.5537	3.36%
63	16.47	18.18	17.32	1.2048	6.95%	1.5096	8.71%
64	19.85	17.01	18.43	2.0069	10.89%	2.5146	13.64%
65	33.59	30.37	31.98	2.2714	7.10%	2.8461	8.90%
66	51.05	47.12	49.09	2.7792	5.66%	3.4824	7.09%
67	50.32	47.82	49.07	1.7677	3.60%	2.2149	4.51%
68	14.83	15.61	15.22	0.5522	3.63%	0.6919	4.55%
69	38.37	35.99	37.18	1.6858	4.53%	2.1123	5.68%
70	28.13	24.42	26.27	2.6237	9.99%	3.2874	12.51%
72	23.81	21.83	22.82	1.4007	6.14%	1.7551	7.69%
73	5.99	6.66	6.33	0.4673	7.39%	0.5856	9.26%
74	33.47	34.50	33.99	0.7318	2.15%	0.9170	2.70%
76	29.21	32.41	30.81	2.2600	7.34%	2.8318	9.19%
77	36.14	34.74	35.44	0.9854	2.78%	1.2347	3.48%
78	10.16	13.64	11.90	2.4624	20.69%	3.0854	25.92%
79	12.01	13.31	12.66	0.9133	7.21%	1.1444	9.04%
80	16.25	14.64	15.44	1.1371	7.36%	1.4248	9.23%
82	22.77	22.26	22.51	0.3585	1.59%	0.4492	2.00%
83	23.20	21.26	22.23	1.3715	6.17%	1.7185	7.73%
84	10.69	7.99	9.34	1.9106	20.46%	2.3940	25.64%
85	11.71	11.56	11.64	0.1065	0.92%	0.1335	1.15%
86	15.33	14.61	14.97	0.5094	3.40%	0.6383	4.26%
87	8.52	8.08	8.30	0.3151	3.80%	0.3949	4.76%
88	6.86	7.36	7.11	0.3524	4.96%	0.4415	6.21%
89	11.36	9.32	10.34	1.4412	13.94%	1.8059	17.47%
90	10.37	9.85	10.11	0.3680	3.64%	0.4611	4.56%
91	14.21	13.39	13.80	0.5781	4.19%	0.7244	5.25%
92	10.90	10.08	10.49	0.5753	5.48%	0.7209	6.87%
93	8.03	8.19	8.11	0.1152	1.42%	0.1443	1.78%
94	19.76	18.09	18.92	1.1786	6.23%	1.4767	7.80%
95	19.21	18.24	18.72	0.6898	3.68%	0.8643	4.62%
96	4.56	4.38	4.47	0.1204	2.69%	0.1508	3.37%
97	9.34	9.74	9.54	0.2813	2.95%	0.3525	3.69%
98	10.83	10.21	10.52	0.4356	4.14%	0.5457	5.19%
99	14.48	15.02	14.75	0.3756	2.55%	0.4706	3.19%
100	16.88	17.75	17.31	0.6121	3.54%	0.7669	4.43%
101	30.16	31.04	30.60	0.6258	2.05%	0.7842	2.56%
102	22.19	21.40	21.80	0.5568	2.55%	0.6977	3.20%

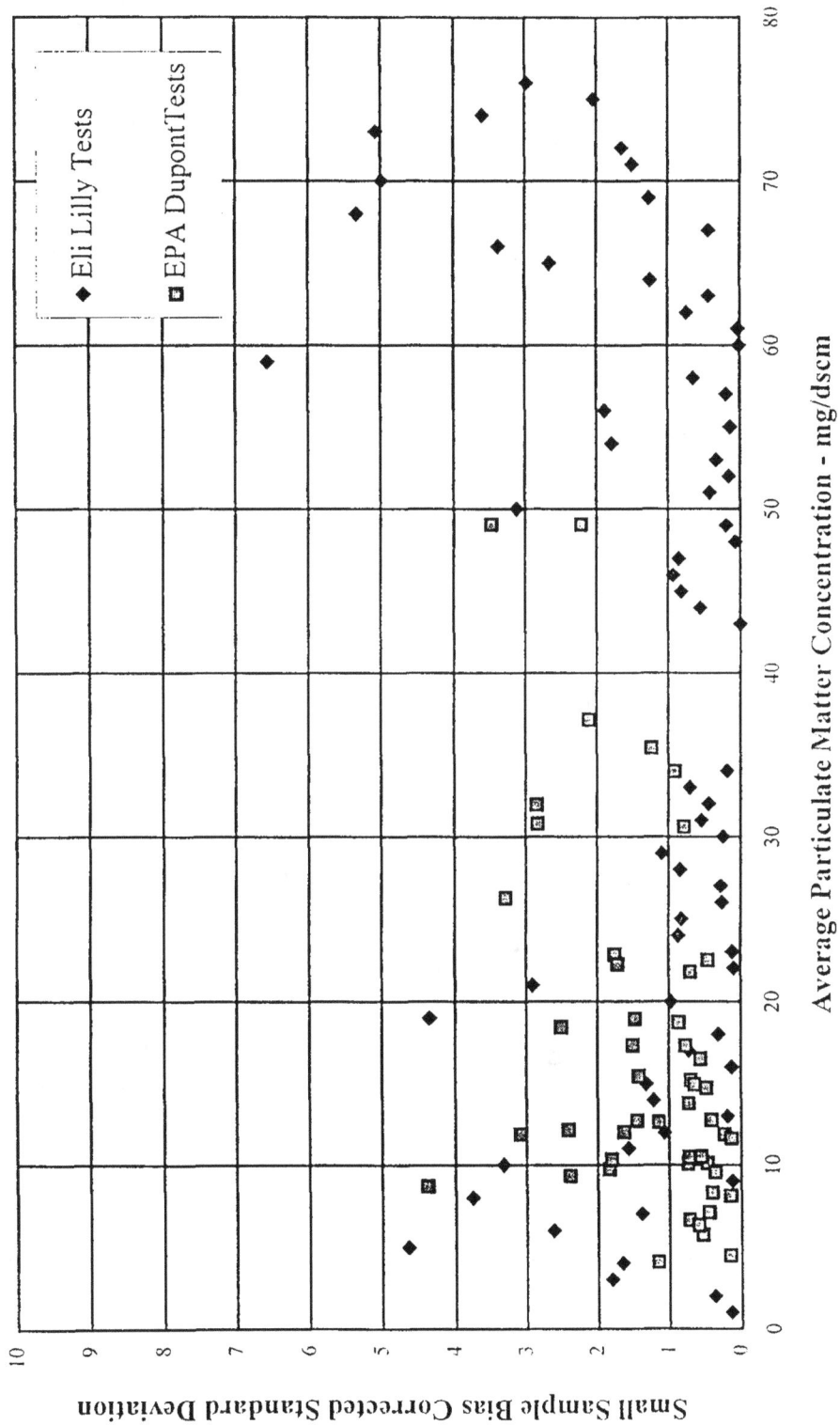

Figure 11a. EPA Method 5i Data - Standard Deviation

Figure 11b. EPA Method 5i Data - Relative Standard Deviation

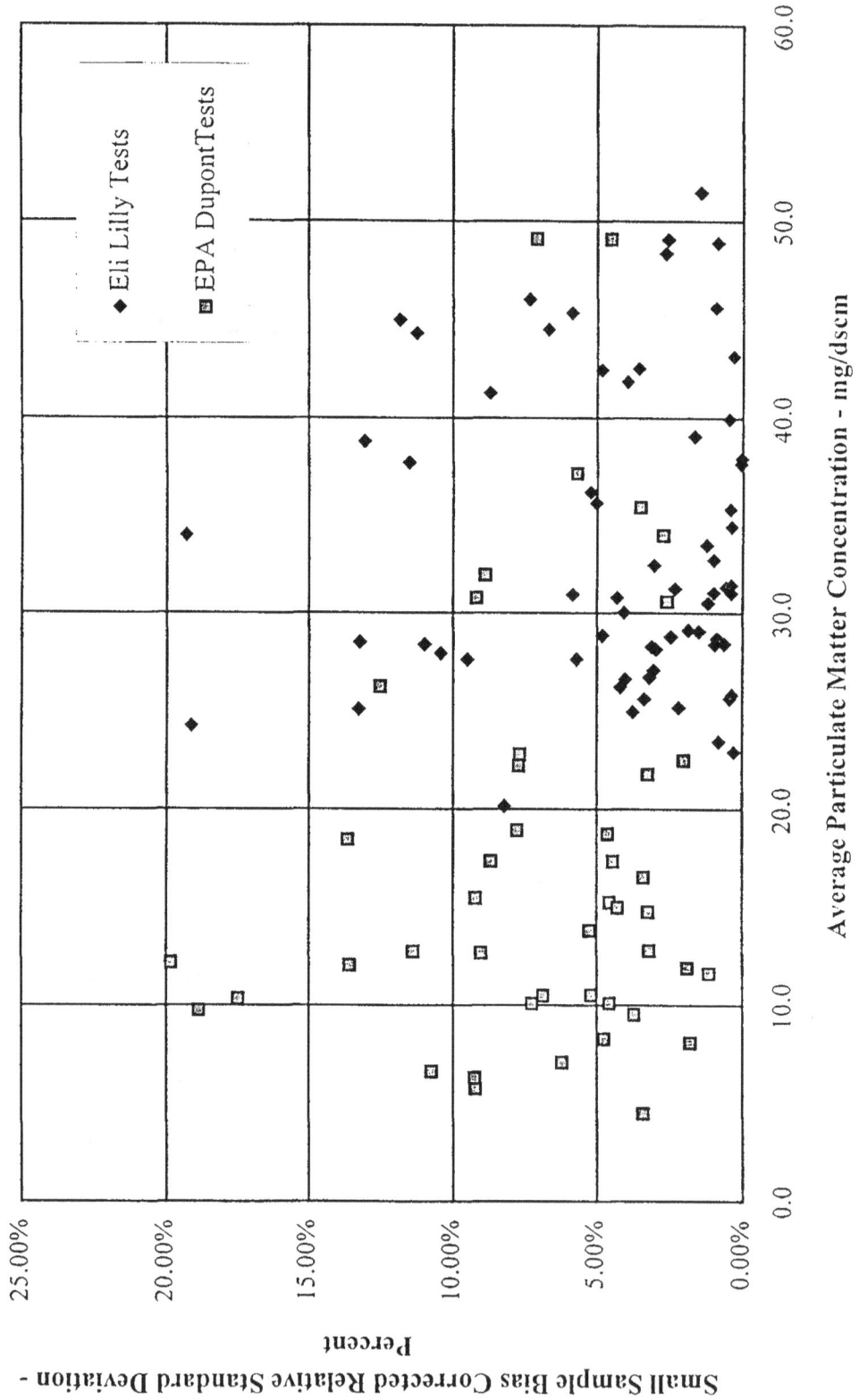

The screened and small sample bias corrected data have been subjected to regression analysis. Since all Method 5i data were obtained using dual trains, no weighting of the data is required. Results from the regression analysis are presented in Table 10 below.

Table 10.   Results of Regression Analysis for Method 5i Data

Parameter	Analysis Result
p	0.2432
Ln(k)	-0.9943
k	0.370
Standard error for p	0.1848
Degrees of freedom	112
Sum of Residuals	-2.101 E-14
Weighted Sum of Squares of Residuals	144.3214
T	1.32
Standard Deviation of Residuals	1.1351

The regression analysis indicates that the estimated standard deviation varies with concentration according to the relationship

$$S = 0.607 \, C^{0.243}$$ Eq 7.

This equation includes the log-log transformation bias correction term, which for the current data set is 1.640. It is critically important to note that the t-parameter is only 1.32, which is well below the critical value of the t-statistic. Thus, based on available data, there is no statistically significant relation between standard deviation for Method 5i and particulate concentration. Based on the regression analysis, the value of the power term (at the 95% confidence level) is:

$$P = 0.243 \pm 0.366 \text{ or from } -0.123 \text{ to } 0.609 [4]$$

---

[4] The predicted value of p (0.243) is taken directly from the regression analysis. The ± 0.3659 term is the product of the t-statistic for 114 degrees of freedom (1.980) and the standard error for p (0.1848).

This finding is consistent with the earlier observation (see Figure 11a) that the standard deviation data appears to be broadly distributed in a box covering the full range of concentration and standard deviation between zero and about 6 mg/dscm.

It is mathematically possible to construct a range of method precision metric for Method 5i as a function of concentration but the indicated relations would have little statistical significance. In such situations, the analysis approach is to calculate a pooled standard deviation for the data. The analysis procedure is described in the Appendix but essentially involves calculation of the weighted average of the variance for the available data. Variance is equal to the square of the standard deviation and the weighting factor for each data point is the number of degrees of freedom for the data point. Since all available Method 5i data are from paired train tests (DF=1) all weighting factors are 1.0 and thus the pooled standard deviation is simply the square root of the sum of the squares for the S_i values. As explained in the Appendix, the appropriate values of S_i to calculate the pooled standard deviation are taken directly from the raw data without adjustment for small sample bias.

The above described pooling procedures were applied to the data in Table 9a and b. There are 114 individual data pairs and the sum of the individual variances is 234.13. Accordingly the pooled variance is 2.054 (234.13/114) and the pooled standard deviation is 1.433 mg/dscm. This single value is the best estimate available for the Estimated σ for Method 5i.

Table 1 in the Appendices provides a list of factors to calculate the confidence interval (at the 95% confidence level) for a pooled value of Est. σ. Using linear interpolation, the values of $P_{0.025}$ and $P_{0.975}$ are 0.883 and 1.155 respectively. Accordingly, the upper and lower 95% confidence intervals on Estimated σ are 1.265 and 1.655 mg/dscm respectively. The confidence intervals are also constant values - not a function of concentration. Figure 12 presents a scatter plot of the Method 5i data along with the estimated values of standard deviation and the confidence intervals. Since Est. σ and the confidence intervals are constants, they are illustrated as straight lines in Figure 12.

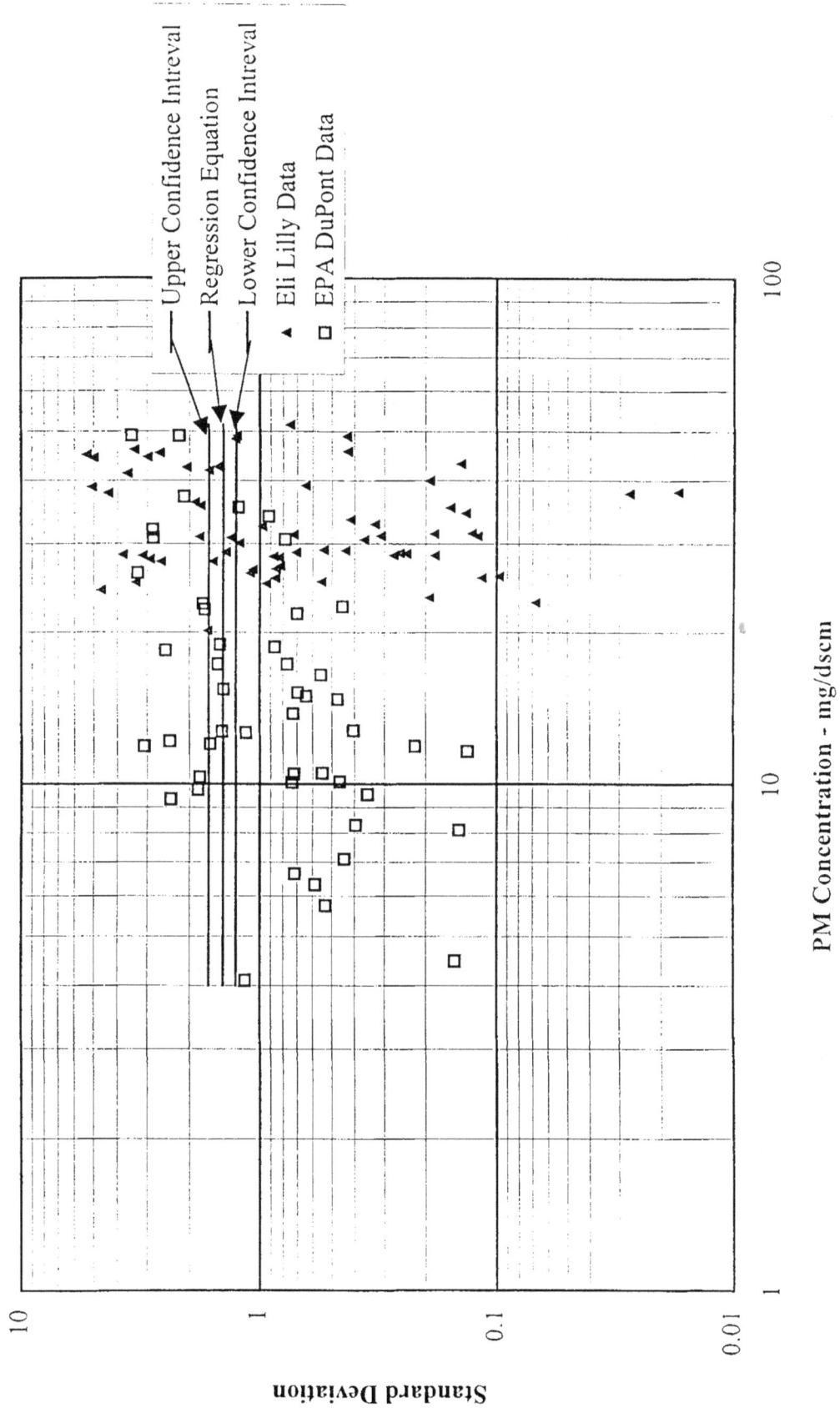

Figure 12. Regression Line and 95% Confidence Interval for Method 5i

PM Concentration - mg/dscm

Standard Deviation

Upper Confidence Intreval
Regression Equation
Lower Confidence Intreval

▲ Eli Lilly Data
□ EPA DuPont Data

60

The estimate of standard deviation provides information on the anticipated range of future measurements using Method 5i. Figure 13 presents three precision metrics for Method 5i. Included are the relative standard deviation, the expected bounds for 99 out of 100 future individual measurements as well as the anticipated bounds for the average of triplicate measurements. These precision metrics have been normalized by the concentration. Thus, even though Est. σ is a constant, normalized precision metrics are strong functions of concentration.

The general presentation approach adopted for the current report is to present a figure defining the anticipated range of concentrations for 99 out of 100 future measurements, under three different scenarios for an assumed variation of standard deviation. Such a figure provides little information under conditions where the standard deviation is evaluated to be essentially a constant. It is much cleaner to simply state the anticipated variation in future measurements.

- If the true standard deviation is essentially equal to the pooled standard deviation, then 99 out of 100 future Method 5i measurements are anticipated to fall with ± 3.68 mg/dscm of the true concentration. This assumes that there is no bias in the measurements and that the true concentration is between about 4 and 50 mg/dscm.
- If the true standard deviation for Method 5i is essentially equal to the lower 95% confidence interval, then 99 out of 100 future measurements are anticipated to fall with ± 3.25 mg/dscm of the true concentration.
- If the true standard deviation for Method 5i is essentially equal to the upper 95% confidence interval, then 99 out of 100 future measurements are anticipated to fall with ± 4.25 mg/dscm of the true concentration.

4.3   Discussion of Particulate Matter Measurement Results

The forgoing discussion provides strikingly different conclusions relative to the precision of Methods 5 and 5i. Specifically, the regression analysis indicates that the standard deviation of Method 5 is a strong function of concentration. In fact, it is suggested that a reasonable interpretation of the data is that Method 5 has a constant relative standard deviation. In contrast,

Figure 13. Method 5i Precision Metrics

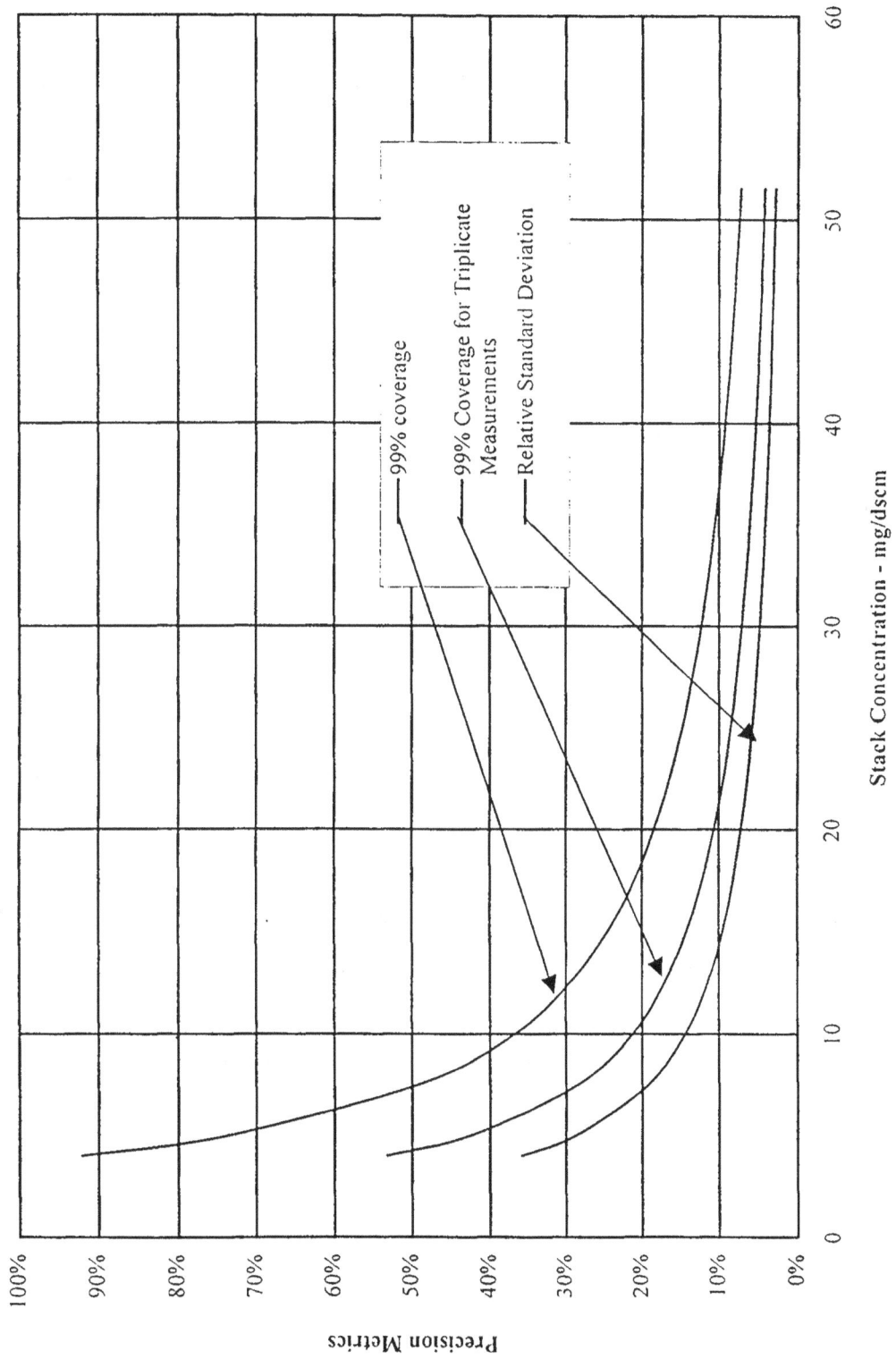

analysis of Method 5i data could detect no relationship between standard deviation and concentration at the 95% confidence level.

Prior discussion has suggested that random error associated with the sample collection process tends to drive the power function term (p) in the regression equation toward 1.0. Similarly, random error in the analytical process tends to drive p toward zero. These are only anticipated trends but they do suggest that different types of random error have driven the assessment of these two particulate measurement methods.

When Method 5 is applied at high particulate concentration, it is reasonable to anticipate that the filter weighing process is sufficiently precise that it contributes negligibly to the overall precision of the method. A large portion of the multi-train Method 5 data was collected from high concentration stacks (>90 mg/dscm). Further, the high concentration data were collected during a time frame when many sampling teams were gaining experience with application of the method. Thus, results from the statistical analysis of Method 5 data can easily be rationalized.

The Method 5i data were all collected under low particulate concentration conditions (<50 mg/dscm with the majority of the data at much lower concentration). Under these conditions, it is anticipated that imprecision of the weighing process may contribute significantly to the overall Method's precision. As described earlier, collected samples must be dried before recording the final particulate weight gain. Collected samples are placed in a dessicator and repeatedly weighed until the tare weight is stabilized. The sample is considered to have reached its final weight if the repeated weighings[5] agree within ±0.5 mg or ±1.0% of the tare weight, whichever is greater. For a typical Method 5 or 5i particulate measurement, a sample is collected from the stack for approximately one hour during which time approximately 1 cubic meter of flue gas is extracted. Thus the process of determining particulate loading on the filter is no more precise than ±0.5 mg and the concentration measurement is no more precise than ±0.5 mg/dscm. That component of measurement imprecision is insignificant when the stack concentration is on the order of 100 mg/dscm. However, for measurements in stacks with PM concentrations on the order of 10 mg/dscm, this represents a significant relative contribution to overall measurement precision.

---

[5] Samples must remain in a dessicator for a minimum of 6 hours between weighings.

Based on the above considerations it is reasonable to anticipate that the analytical portion of Method 5i measurements will contribute significantly to the overall precision of the method and that random error in the weighing process will not vary significantly with particulate concentration. Method 5i is also subject to random error associated with sample collection and the influence of those errors on method precision should vary with concentration. In that regard, it is important to note that all available Method 5i data have been collected with dual train measurements. Further, the measurement teams expended significant effort developing the skills of the field technicians. In fact, data from numerous tests were discarded as a result of personnel climbing the learning curve. Finally, all data in the Method 5i data set were screened to eliminate any data points with RSD above about 15%. All of these factors contribute to the observation that random error in the sampling process has a small contribution to the indicated overall precision of Method 5i.

The above considerations are provided as one possible rationalization for the observed differences in the precision characteristics of the two methods. These factors reflect on both the characteristics of the measurement methods and the characteristics of the available data used to assess the methods. As regards Method 5, it appears likely that the relative experience of sampling teams collecting data in the low and high concentration ranges has impacted the characteristics of data used to assess the method. Further it is likely that the true dependence of Method 5 standard deviation on concentration is less than indicated by the current data set.

As regards Method 5i, recall that regression analysis of the data does not deny the possibility that σ is a function of particulate concentration. The analysis merely concludes that no relationship can be confirmed at the 95% confidence level. For that reason a pooled analysis of the data was performed. Considering the physical similarities between the hardware and analytical procedures for Methods 5 and 5i, it is likely that standard deviation for Method 5i increases (at least slightly) with increasing stack particulate concentration.

It is important to place the above considerations into practical perspective. When Method 5i is used to measure PM loading in low particulate concentration stacks (<50 mg/dscm), analysis of available data suggests that 99 out of 100 future measurements should fall within about 3.7 mg/dscm of the true concentration. The original data set (before outlier screening) contained 116 pairs of dual train results. Assuming that the true concentration for each test is the average of the test pair, dividing the spread in each pair by 2 provides a crude way to assess this prediction. For the entire data set (Tables 10a and 10b) the maximum spread between any data pair occurred in run number 53 from the Eli Lilly data set. The reported concentrations for that data pair were 30.3 and 37.7 with an average concentration of 34.0 mg/dscm. Thus, for 116 paired measurements (232 applications of the Method) the maximum difference between the measured concentration and the estimated true concentration was 3.7 mg/dscm. Interestingly, that measurement was determined to be an outlier and was eliminated from the overall analysis.

It is certainly possible (even likely) that the precision of Method 5i has some dependence on concentration. However, the forgoing analysis suggests that the constant sigma assumption provides a reasonable and practical estimate of the Method's precision.

It is unlikely that the imprecision of Method 5 increases as rapidly with concentration as suggested by the data in Figures 7, 8 and 9. It is expected that increased experience of field sampling teams, including the lessons learned from application of Method 5i, have already reduced the range of random errors impacting Method 5 results. It is anticipated that the inherent precision of Method 5 should be similar to that for Method 5i but σ for the method almost certainly does increase with concentration. Note that the Method 5 data from the Pittsfield tests were collected in the mid-1990s and should reflect increased experience for the testing team. Data from this test series (16 valid data pairs) were examined to determine the range of the data pairs relative to the pair average. For runs 15 and 17, the spread minus the mean were 4.4 and 6.8 mg/dscm respectively while the spreads for the remainder of the data were less than 3.5 mg/dscm. This suggests that Method 5 might provide slightly less precise results than Method 5i but not dramatically less. Even this observation must be tempered by the fact that the standard deviation for the method increases with increasing concentration. Based on these qualitative considerations, it is suggested that the data in Figures 7, 8, and 9 should be taken as an upper limit on the imprecision of Method 5. To fully assess Method 5 precision at higher concentrations (>100mg/dscm) additional multi train data is required in that concentration range.

This page Intentionally Left Blank

5.0   EPA Method 23 For Measuring Dioxin and Furan Emissions

The EPA method for measurement of dioxin and furan stack emissions is denoted as Method 23 (56 FR 67788 and 40 CFR Part 60 – Appendix A). The hardware for the method is illustrated in Figure 14 and has several similarities to the hardware discussed previously for Method 5. The major difference is addition of a module filled with an absorbent material known as XAD. A small circulating pump maintains the temperature of the XAD module at approximately 60° F. The XAD module is spiked with known quantities of labeled compounds that are used for both system calibration and to experimentally determine the recovery of the overall sampling and analysis process.

The quantity of dioxin and furan collected and analyzed by the method is extremely small. Typical stack concentrations are on the order of a few ng/dscm. To collect sufficient material for analysis by high resolution GC/MS, each sampling run extends for at least three hours and may last for more than 6 hours. There are eight possible homologues of both polychlorinated dibenzo(p)dioxin and polychlorinated dibenzofuran. Only those homologues with four to eight chlorine atoms are believed to have adverse health effects. Accordingly, some environmental regulations (i.e., the rules governing municipal waste combustors) limit the release of all tetra through octa-chlorinated dioxins and furans. Other regulations take account for the fact that different dioxin and furan congeners have vastly different toxicity. Most of the world has adopted a listing of relative congener toxicity that was developed under the auspices of NATO. These relative toxicity factors are multiplied by the concentration of each congener to yield an emission concentration that is equivalent to the toxicity that would occur if all of the indicated mass was found as the most toxic congener (i.e., 2,3,7,8 tetrachloro dibenzo(p)dioxin). Emissions expressed in this manner are referred to as International Toxic Equivalent, or ITEQ. Regardless of whether an emission standard is expressed as total mass of tetra through octa or as ITEQ, the sampling and analysis procedure defined by Method 23 is the same. The only difference is that the ITEQ process provides significant weighting factors to a select group of individual congeners. As will be shown in material that follows, application of these weighting factors does impact the indicated precision of the measurement method.

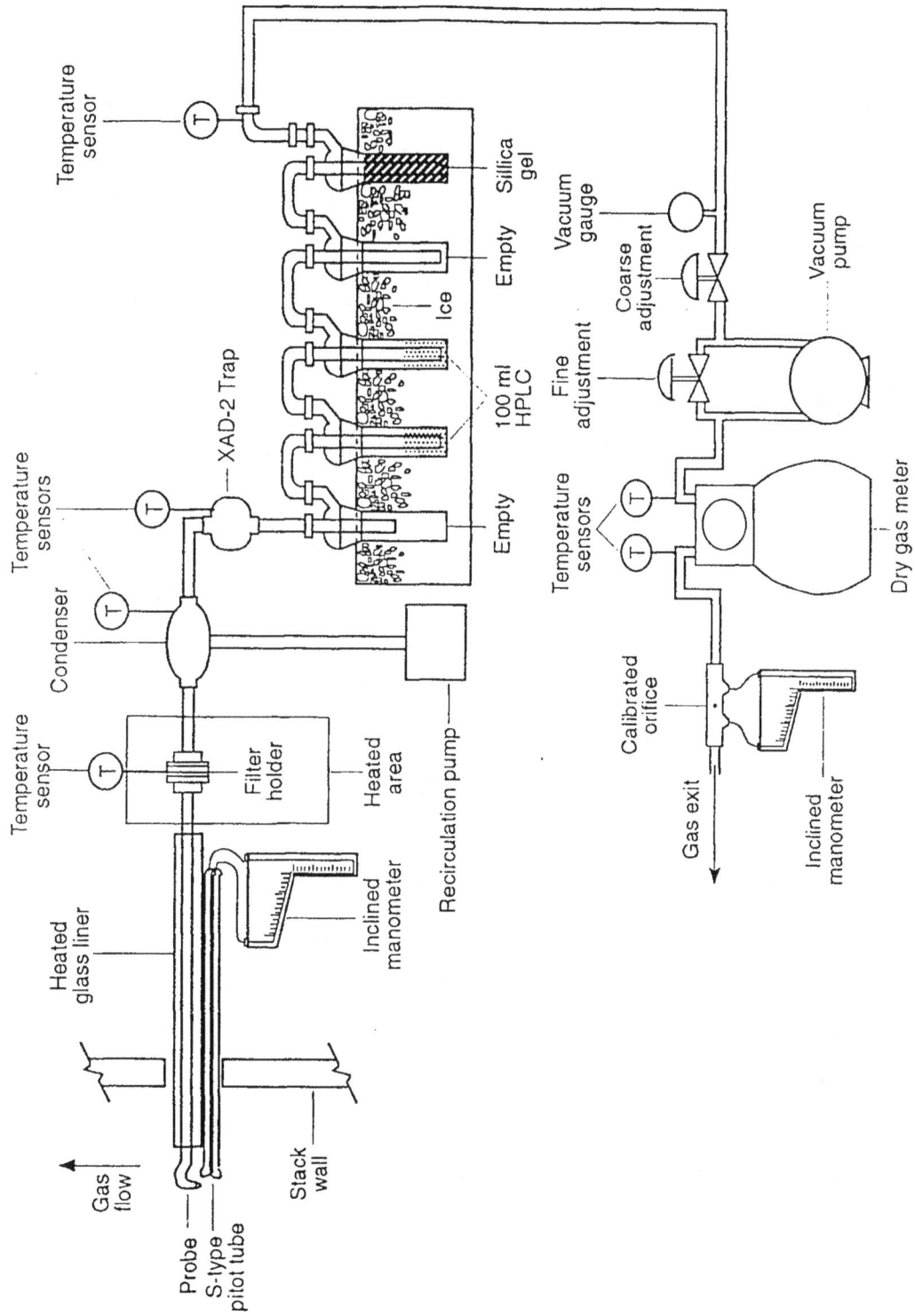

Figure 14. Schematic of Method 23 sampling train.

68

5.1 Available Multi-Train Data for Method 23 as Total PCDD/PCDF

Extensive effort has been expended in development of sampling and analytical methodology for determination of dioxin and furan emission concentrations. Separate procedures have evolved in Europe, Canada and the US which are similar in many respects. There are however, small differences in the analytical procedures that may have significant impact on the precision of measurement results. Accordingly, the current study focuses only on US EPA Method 23 since that is the method which must be used to determine compliance with US emission standards. Before presenting the available multi-train data for Method 23, it is important to note that the procedures used by EPA to validate Method 23 are different from those used to validate other methods. Specifically, in lieu of gathering multi-train data from one or more source categories, the Agency developed hardware and a procedure for dynamically spiking a sampling train with known quantities of isotopically labeled dioxin and furan congeners. Validation of the method focused on experiments determining the fractional recovery of the dynamically spiked compounds.

There are a variety of approaches that may be used to validate performance of a method. Dynamically spiking a sampling train with a known quantity of a tracer compound is a potentially valid approach. In fact many of the EPA method validation efforts have used dynamic spiking in quad train tests to gain information on both precision and bias of an emerging method. However, use of dynamic spiking experiments as the sole approach for method validation is valid only under conditions where there is no possibility for formation of the pollutant of interest in the sampling train itself (e.g., for methods measuring the total emission of an element such as a heavy metal). As regards dioxin and furan measurement, there is a significant potential for formation of these compounds under thermal conditions that may occur within a sampling probe or within the hot filter box of the Method. The potential for formation of the target analytes within the sampling train raises serious concern about the completeness of EPA studies validating Method 23.

The report describing the Method 23 validation effort indicates that a quad-train was dynamically spiked with isotopically labeled dioxin and furan congeners (MRI, 1991). The report indicates that the collected samples were analyzed for both the native congeners as well as the dynamically spiked congeners. In fact, the report includes tables listing the collected mass of native dioxin and furan in each homologue group as well as the collected mass of spiked compounds. Unfortunately, no documentation was found describing the volume of sample gas collected by each train or the thermal conditions in the trains. Thus, insufficient data are provided to calculate or estimate the concentration of PCDD/PCDF in those samples. The method validation appears to be based on comparison of the relative mass of spiked compounds in each train.

The above comments do not mean that EPA Method 23 is a technically unacceptable method for determining exhaust concentrations of dioxin and furan. It is asserted, however, that validation of the Method is not complete. Nonetheless, based on the similarity of Method 23 with procedures used in other countries, it is anticipated that Method 23 is at least as good as those used in Europe and Canada.

Lacking multi-train data from EPA method validation testing, multi-train data for EPA Method 23 is very rare. The only major set of PCDD/PCDF multi-train data uncovered in the current study is the data provided by the tests of Rigo and Chandler on a municipal waste combustor in Pittsfield Massachusetts. The Pittsfield data were gathered at a mid-point in the air pollution control system for the facility. Accordingly, results presented below do not represent stack emission levels for this facility. Additionally, there are three data points collected from a lightweight aggregate kiln. These tests are significant in that two different sampling companies collected the paired data simultaneously, from the same stack. One testing group (EER,1997) gathered data for an EPA sponsored study while a second contractor (TRC) was gathering confirmation measurements (sponsored by the host facility). Table 11 provides a listing of available data.

5.2   Analysis of Method 23 Data for Total Dioxin and Furan.

Data in Table 11 for Total Dioxin and Furan (mass of tetra through octa homologues) are presented in Figures 15 and 16. Figure 15 shows the variation of standard deviation as a function of concentration (average of the test pair), while Figure 16 shows the same information in a plot of relative standard deviation versus concentration. Note that the range of these data is extremely limited and that the measured RSD values are quite large, especially at lower concentrations. Expressed as total mass, the data range is from less than 1 to about 27 ng/dscm.

Table 11.   Method 23 Data as Total Mass of Tetra through Octa Dioxin plus Furan

Data Source	Run Number	Train A	Train B	Average Concentration	Standard Deviation	RSD	S - Bias Corrected	RSD - Bias Corrected
Rigo & Chandler	1	5.97	5.05	5.51	0.6456	11.72%	0.8090	14.68%
	2	2.11	2.26	2.18	0.1031	4.72%	0.1291	5.91%
	3	3.00	2.70	2.85	0.2179	7.65%	0.2731	9.58%
	4	1.14	1.56	1.35	0.2955	21.86%	0.3703	27.39%
	5	2.93	3.61	3.27	0.4806	14.71%	0.6022	18.43%
	6	0.84	0.94	0.89	0.0718	8.03%	0.0900	10.06%
	7	13.09	17.56	15.33	3.1618	20.63%	3.9617	25.85%
	8	20.17	20.23	20.20	0.0445	0.22%	0.0558	0.28%
	9	11.11	8.86	9.99	1.5964	15.99%	2.0002	20.03%
	10	8.31	8.12	8.22	0.1364	1.66%	0.1709	2.08%
	11	5.40	5.27	5.33	0.0972	1.82%	0.1218	2.28%
	12	2.52	1.44	1.98	0.7625	38.57%	0.9554	48.32%
	13	1.51	1.60	1.55	0.0674	4.34%	0.0845	5.43%
	14	0.87	0.78	0.83	0.0645	7.78%	0.0808	9.75%
	15	1.24	0.88	1.06	0.2544	24.07%	0.3188	30.16%
	16	1.05	0.88	0.96	0.1180	12.24%	0.1478	15.33%
	17	0.63	1.46	1.05	0.5872	56.14%	0.7357	70.34%
	18	0.69	0.72	0.70	0.0227	3.22%	0.0284	4.04%
	19	0.55	0.53	0.54	0.0137	2.53%	0.0172	3.17%
EER-TRC	1	5.82	3.69	4.76	1.5039	31.62%	1.8844	39.62%
	2	15.34	16.22	15.78	0.6232	3.95%	0.7809	4.95%
	3	27.72	27.39	27.55	0.2349	0.85%	0.2943	1.07%

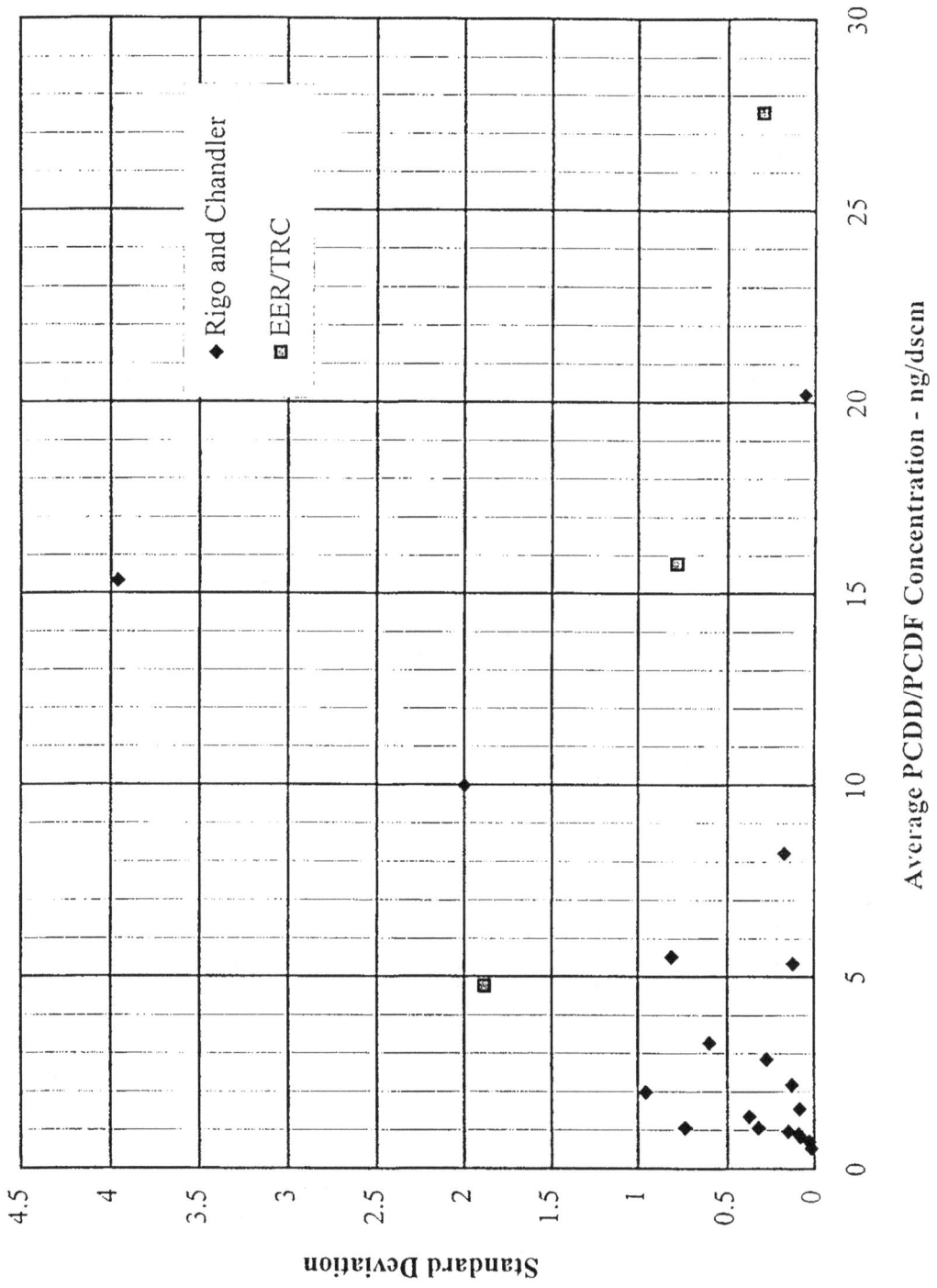

Figure 15. Method 23 Data - Total PCDD/PCDF
Standard Deviation

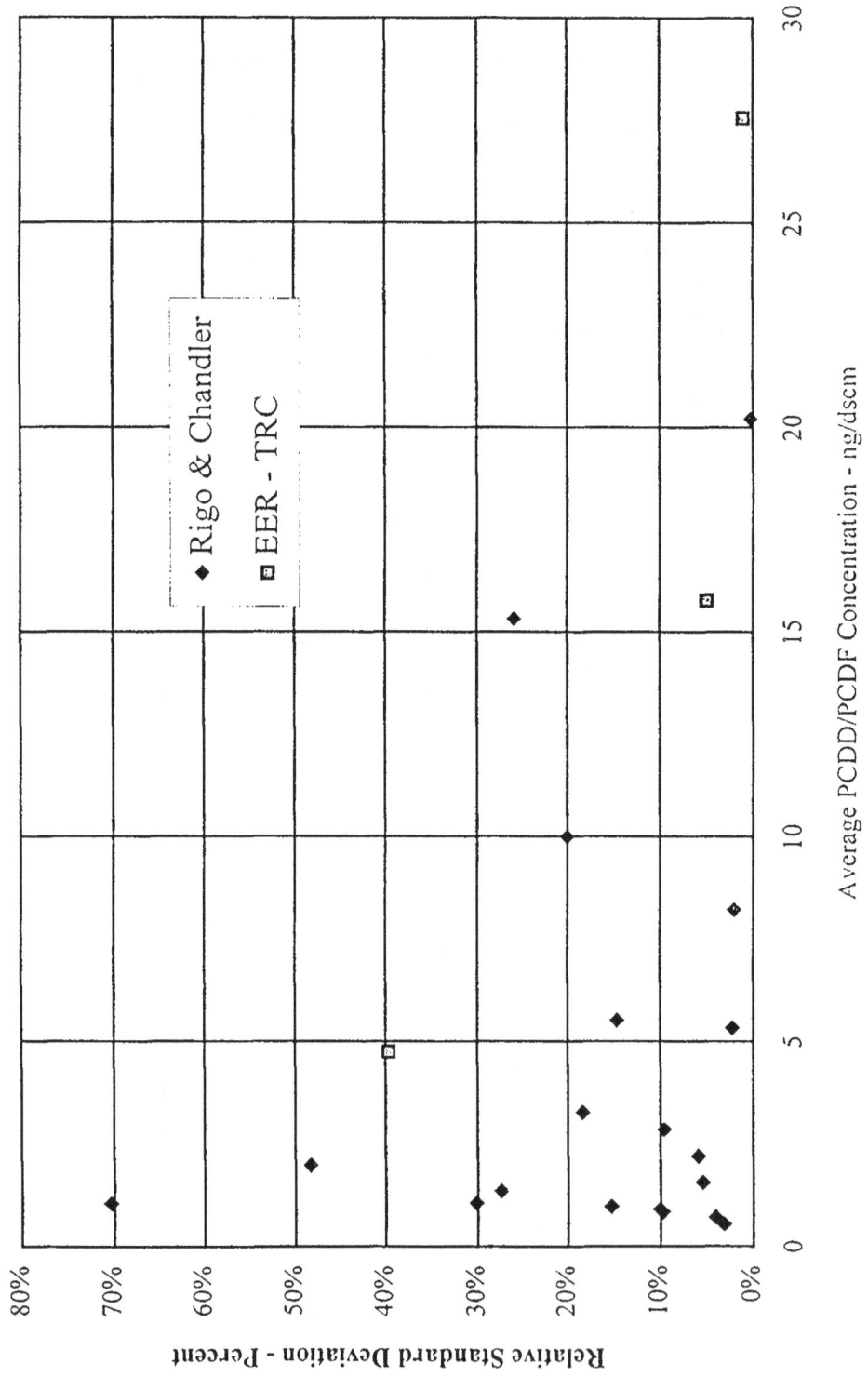

Figure 16. EPA Method 23 Data - Total PCDD-PCDF Relative Standard Deviation

The SPS data outlier procedures were applied to the data in Table 11. Several different range groupings were evaluated and each grouping identified Run number 7 from the Rigo and Chandler tests as an outlier. The basis for this identification is easily seen in Figure 15. However, when presented as relative standard deviation (see Figure 16), Run number 7 is only slightly above the remainder of the data. The SPC outlier procedures are used to identify data points where the span of data appears to be abnormally large. In this instance, the data point has been examined and a decision was made to retain the data for the remainder of the analysis.

Data from Table 11 were submitted to regression analysis to determine the best curve fit. No data weighting were required since all available results are from dual-train testing. Analysis results indicate that the estimated standard deviation varies as a function of total dioxin plus furan concentration according to the equation

$$S \text{ (for Method 23 as total PCDD/PCDF)} = 0.2722 * C^{0.56}.$$

This equation includes the log-log transformation bias correction factor, which for the current data set is 1.893. The t statistic for the regression is 2.433, which is slightly above the critical t statistic for 20 degrees of freedom at 95% confidence level (2.086). This implies that the observed relation between standard deviation and concentration did not occur by chance but that there is considerable uncertainty in the slope of the relation.

More precisely, at the 95% confidence level, the power function term is best expressed as

$$p = 0.56 \pm 0.48 \text{ or } p \text{ lies between 0.08 and 1.04.}$$

Figure 17 presents the regression results including the small sample bias-corrected data, the regression line and the upper and lower 95% confidence intervals. Figure 18 presents the various precision metrics for Method 23. Data included in Figure 18 is based on the predicted relation between Est. σ and C. If that relation is correct and if the true stack concentration is above about 2 ng/dscm (excluding temporal variation), 99 out of 100 triplicate measurements using Method 23 should fall within ± 30% of the true concentration. Source variation will increase those bounds.

Figure 17. Regression Line and 95% Confidence Interval for Method 23

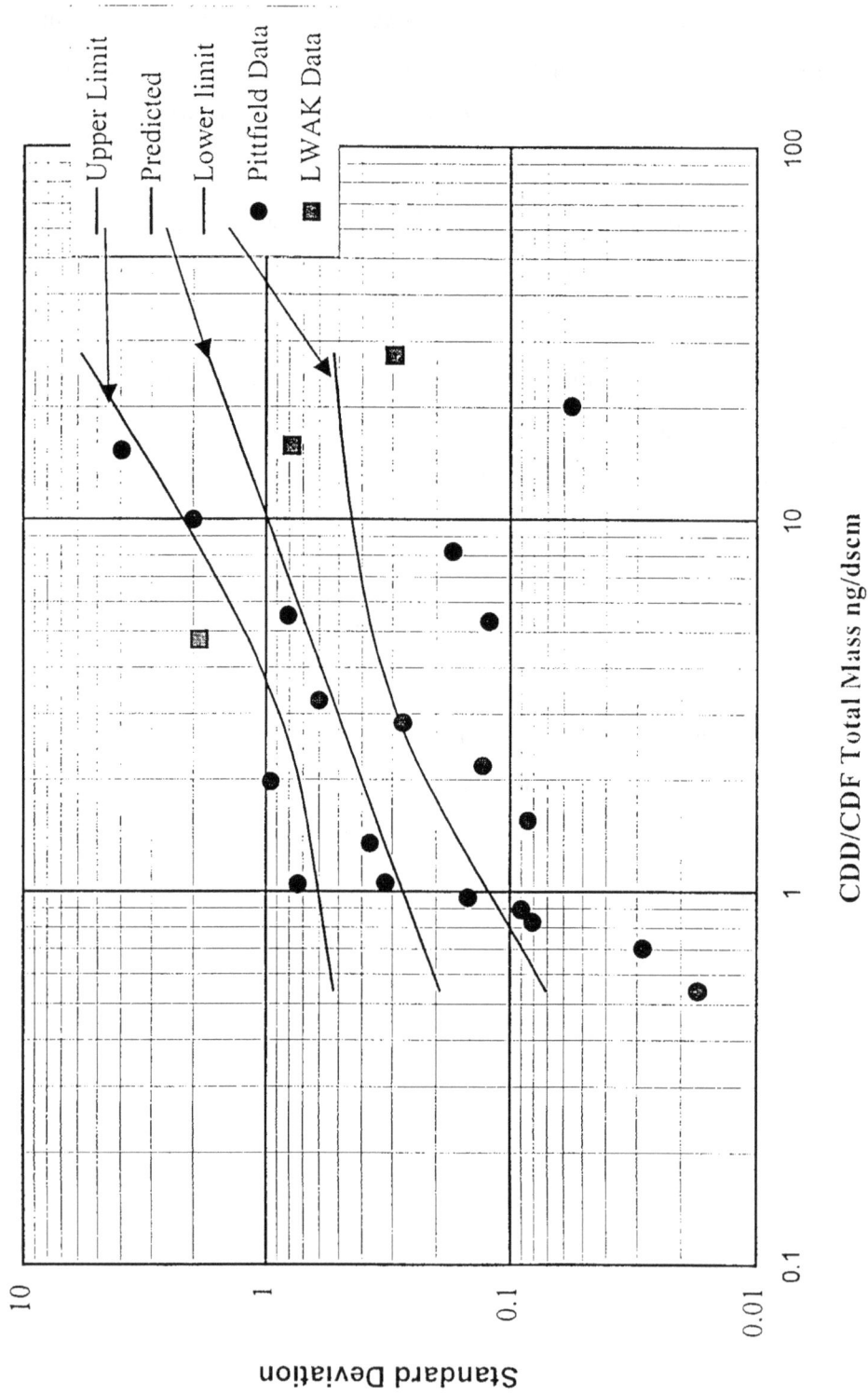

Figure 18. Method 23 Precision Metrics -Total PCDD/PCDF

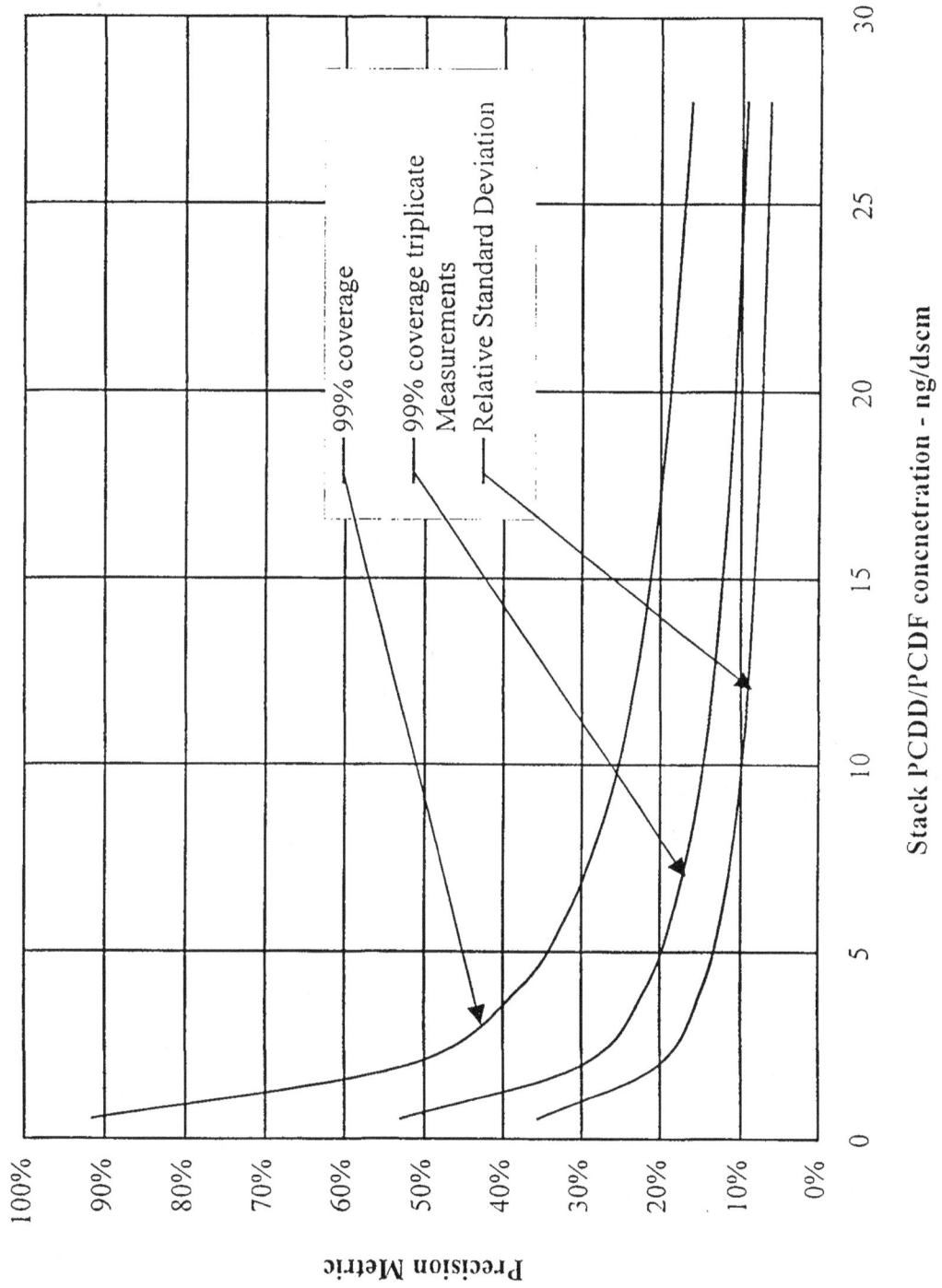

Figure 18. Method 23 Precision Metrics -Total PCDD/PCDF

76

It is critical to recall that there is considerable uncertainty in the value of the slope term in the regression equation. Figure 19 illustrates the anticipated bounds on 99 out of 100 individual measurements for three different scenarios on the S versus C relationship. If the regression relation is the proper description of how standard deviation varies with concentration, the anticipated spread of future data is relatively tight. For example, in that situation, sampling a stack that actually contains 20 ng/dscm of total dioxin plus furan should result in 99 out of 100 measurements falling in the range of 16.3 to 23.7 ng/dscm. Conversely, if the standard deviation of the method is more closely described by the upper confidence interval, a much broader range of data can be anticipated. In this case, sampling the hypothetical stack containing 20 ng/dscm dioxin and furan using Method 23 should yield 99 out of 100 measurements falling in the range of 9.13 to 30.87 ng/dscm.

Table 12 presents a tabular summary of the anticipated range of measured PCDD/PCDF concentration. This table is based on the assumption that the regression equation properly describes the variation of standard deviation with concentration for the best estimate of standard deviation. This is the best estimate available for method precision based on the available data. It is critically important to reiterate that concentration information presented in this table (as well as the concentration data in all the tables and figure in this report) are not corrected to constant excess air level. An example is in order. Assume that a facility operates at 11% O_2 in the stack and must comply with a PCDD/PCDF emission limit of 35 ng/dscm @ 7% O_2. Adjusting the emission standard from 7 to 11 % O_2 shows that the facility must maintain PCDD/PCDF stack concentration below 24.9 ng/dscm. If the true stack concentration was exactly 24.9 ng/dscm, imprecision from repeated application of Method 23 should produce 99 out of 100 measurement results ranging from 20.7 to 29.2 ng/dscm. When converted back to the basis of the standard, the anticipated data range is from 29.1 to 40.9 ng/dscm @ 7% O_2. At the 95% confidence level for this hypothetical facility, a single measurement below 20.7ng/dscm is below the standard while a measurement above 29.2 ng/dscm is above the standard. At the 95% confidence level, results between 20.7 and 29.2 could be either in or out of actual compliance.

The above analysis clearly points to the fact that there is an insufficient body of data available to adequately assess the precision of EPA Method 23. The limited quantity of available data suggests that the precision of Method 23 conforms to the predictions presented in Table 12. However, there is

Figure 19. Precision Estimates for Total PCDD-PCDF Measurements Using EPA Method 23

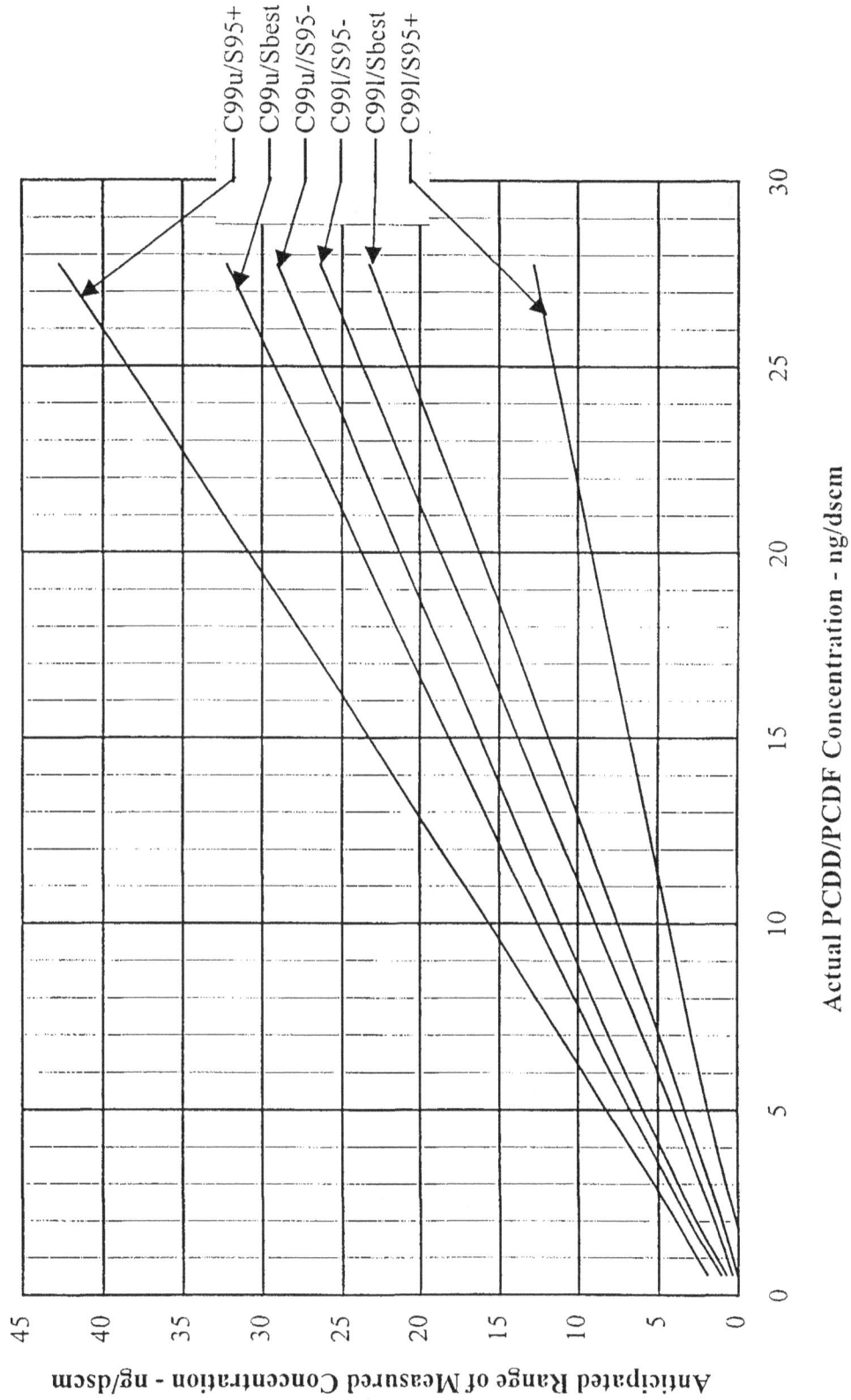

C99u/S95+
C99u/Sbest
C99u/S95-
C99l/S95-
C99l/Sbest
C99l/S95+

Actual PCDD/PCDF Concentration - ng/dscm

Anticipated Range of Measured Concentration - ng/dscm

78

**Table 12. Anticipated Range of Measured PCDD/PCDF Concentration
Based on Best Estimate of Method 23 Standard Deviation**

True Stack PCDD/PCDF Concentration ng/dscm	99 out of 100 Single Measurements		99 out of 100 Triplicate Measurements	
	Lower Limit	Upper Limit	Lower Limit	Upper Limit
0.5	0.97	0.03	0.77	0.23
2.0	3.03	0.97	2.60	1.40
4.0	5.52	2.48	4.88	3.12
6.0	7.91	4.09	7.10	4.90
8.0	10.2	5.76	9.29	6.71
10.0	12.5	7.46	11.5	8.53
12.0	14.8	9.19	13.6	10.4
14.0	17.1	10.9	15.8	12.2
16.0	19.3	12.7	17.9	14.1
18.0	21.5	14.5	20.0	16.0
20.0	23.7	16.3	22.2	17.8
22.0	25.9	18.1	24.3	19.7
24.0	28.1	19.9	26.4	21.6
26.0	30.3	21.7	28.5	23.5
28.0	32.5	23.5	30.6	25.4

significant uncertainty associated with this analysis. If the actual method precision conforms to the upper confidence interval in the analysis, a very wide spread in results can be anticipated. The issue can only be resolved by gathering significantly more data over a broader range of concentrations. Of particular importance would be collection of multi-train data at concentrations above the levels included in the current data set.

5.3   Available Multi-Train Data for Method 23 as ITEQ.

Data for Method 23, with results expressed as ITEQ, are the same as those for Method 23 with results expressed as total mass of tetra through octa dioxin plus furan. Table 13 presents the ITEQ results including the results of each run, the average concentration and the calculated standard deviation from the run. Also included are data following application of the small-sample, bias correction factor. These data are illustrated in Figures 20 and 21 as scatter plots of standard deviation and relative standard deviation versus the average concentration calculated from the data pair. The results are generally similar to those for the Method as total PCDD/PCDF except that the scales are greatly reduced. The range of the concentration data is from about 0.02 to 0.91 ng ITEQ/dscm.

The above Method 23 data were submitted to regression analysis and results indicate that S is related to C according to the equation,

$$S = 0.4795C^{0.345}.$$

Unfortunately the t-statistic on the power term is only 1.02 which is well below the critical value of 2.086. This implies that the regression equation could have occurred by chance and that, at the 95% confidence level, no statistically meaningful relationship between S and C was detected. The reason for this result is easily observable in the data presented in Figure 20. Note that the majority of the data appear as a scattering of points at concentrations below 0.4 ng ITEQ/dscm and standard deviation below 0.04 ng ITEQ/dscm. There are four data points with standard deviation above 0.054 ng ITEQ/dscm. These data might suggest that S increases with increasing C. However, if those four points are discounted, one can easily envision an opposite slope to a regression line. The regression analysis allows the potential range of the power term to be calculated. At the 95% confidence level, the p parameter could be as large as 1.051 and as small as –0.36. The magnitude of the uncertainty concerning the precision of Method 23 for dioxin and furan as ITEQ is further illustrated in Figure 22. This figure shows the small sample bias-corrected data, the regression equation and the upper and lower confidence intervals. It is

Table 13. Method 23 Data as ITEQ

Data Source	Run Number	Train A	Train B	Average Concentration	Standard Deviation	RSD	S - Bias Corrected	RSD - Bias Corrected
Rigo & Chandler	1	0.2847	0.2201	0.2524	0.0457	18.10%	0.0572	22.67%
	2	0.1077	0.1095	0.1086	0.0012	1.15%	0.0016	1.44%
	3	0.1128	0.1384	0.1256	0.0181	14.44%	0.0227	18.09%
	4	0.0584	0.0879	0.0731	0.0208	28.51%	0.0261	35.72%
	5	0.1282	0.1690	0.1486	0.0288	19.42%	0.0361	24.33%
	6	0.0352	0.0499	0.0426	0.0104	24.38%	0.0130	30.55%
	7	0.5678	0.4738	0.5208	0.0664	12.76%	0.0833	15.99%
	8	0.8825	0.9431	0.9128	0.0429	4.70%	0.0537	5.89%
	9	0.4600	0.3779	0.4190	0.0580	13.84%	0.0727	17.34%
	10	0.3812	0.3814	0.3813	0.0002	0.04%	0.0002	0.05%
	11	0.2603	0.2374	0.2489	0.0161	6.48%	0.0202	8.12%
	12	0.0604	0.0411	0.0508	0.0136	26.84%	0.0171	33.64%
	13	0.0382	0.0781	0.0582	0.0283	48.60%	0.0354	60.90%
	14	0.0362	0.0352	0.0357	0.0007	2.09%	0.0009	2.62%
	15	0.0620	0.0297	0.0459	0.0228	49.72%	0.0286	62.30%
	16	0.0474	0.0342	0.0408	0.0094	22.93%	0.0117	28.73%
	17	0.0298	0.0548	0.0423	0.0177	41.89%	0.0222	52.49%
	18	0.0332	0.0303	0.0317	0.0020	6.43%	0.0026	8.06%
	19	0.0217	0.0200	0.0209	0.0012	5.65%	0.0015	7.08%
LWAK	1	0.0555	0.0423	0.0489	0.0093	19.10%	0.0117	23.94%
	2	0.1785	0.1766	0.1776	0.0013	0.74%	0.0016	0.92%
	3	0.3171	0.3193	0.3182	0.0016	0.50%	0.0020	0.62%

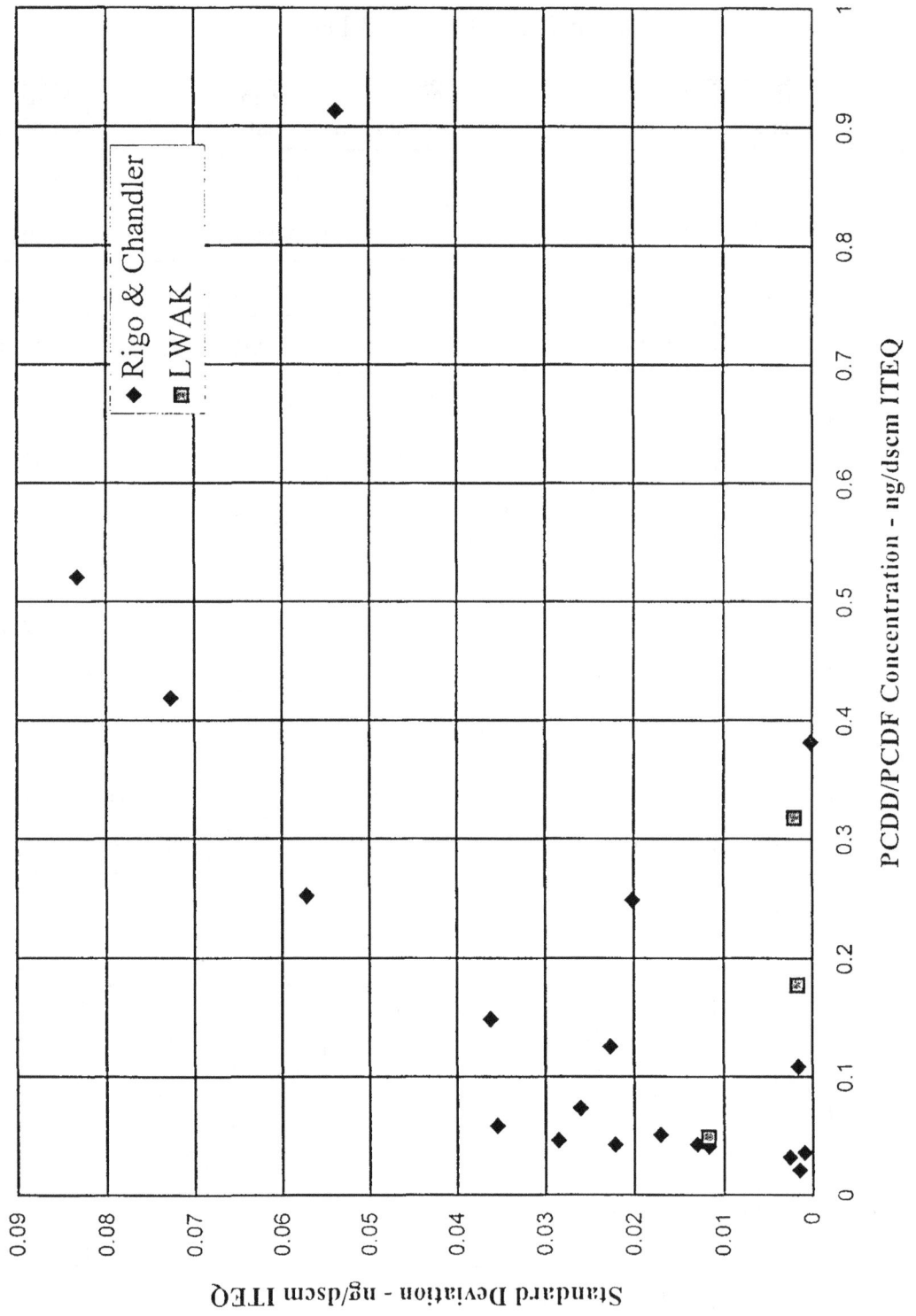

Figure 20. Method 23 Data PCDD/PCDF as ITEQ
Standard Deviation

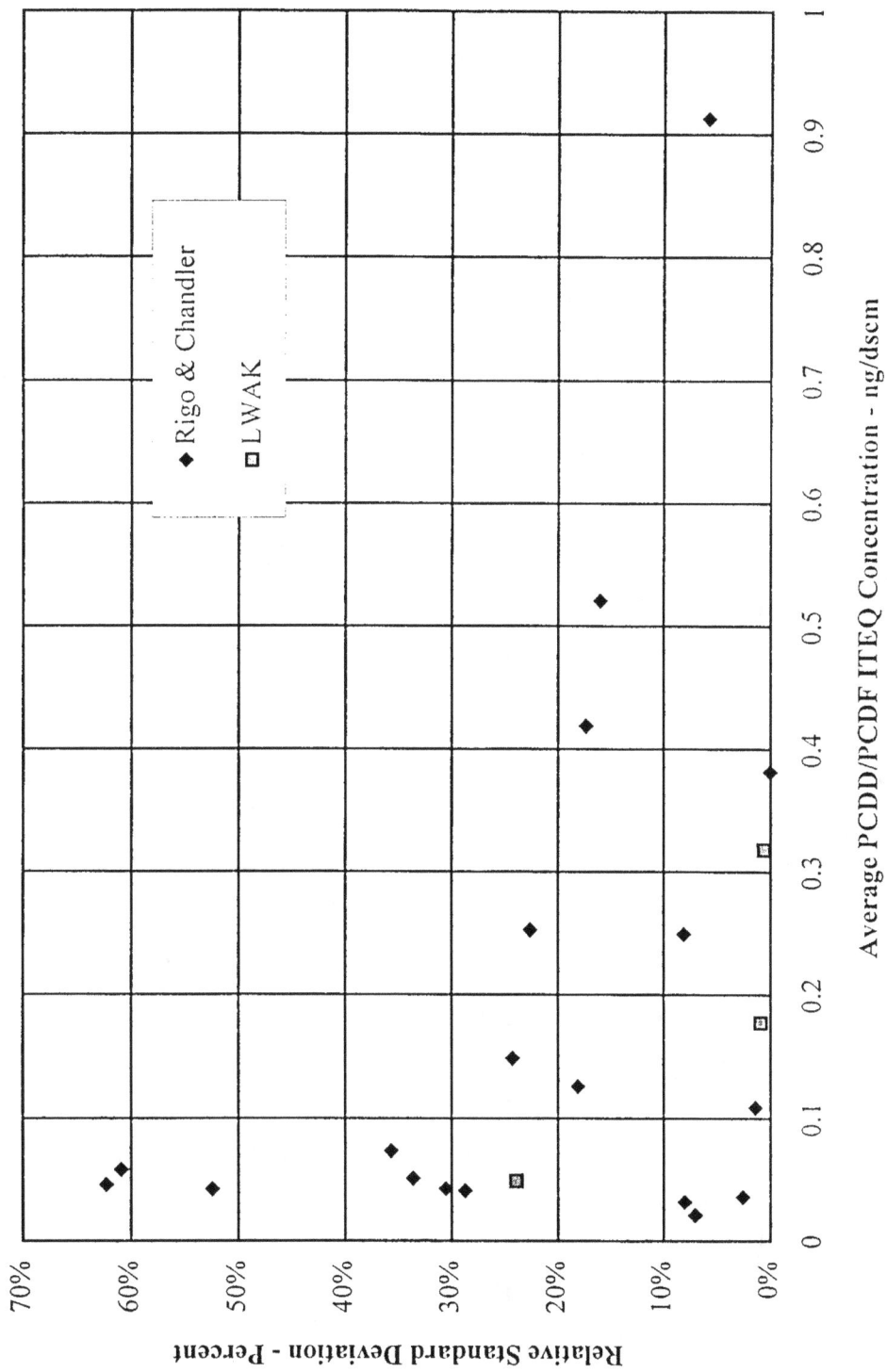

Figure 21. Method 23 Data - PCDD/PCDF as ITEQ
Relative Standard Deviation

Figure 22. Regression Line and 95% Confidence Interval for
EPA Method 23 - PCDD/PCDF as ITEQ

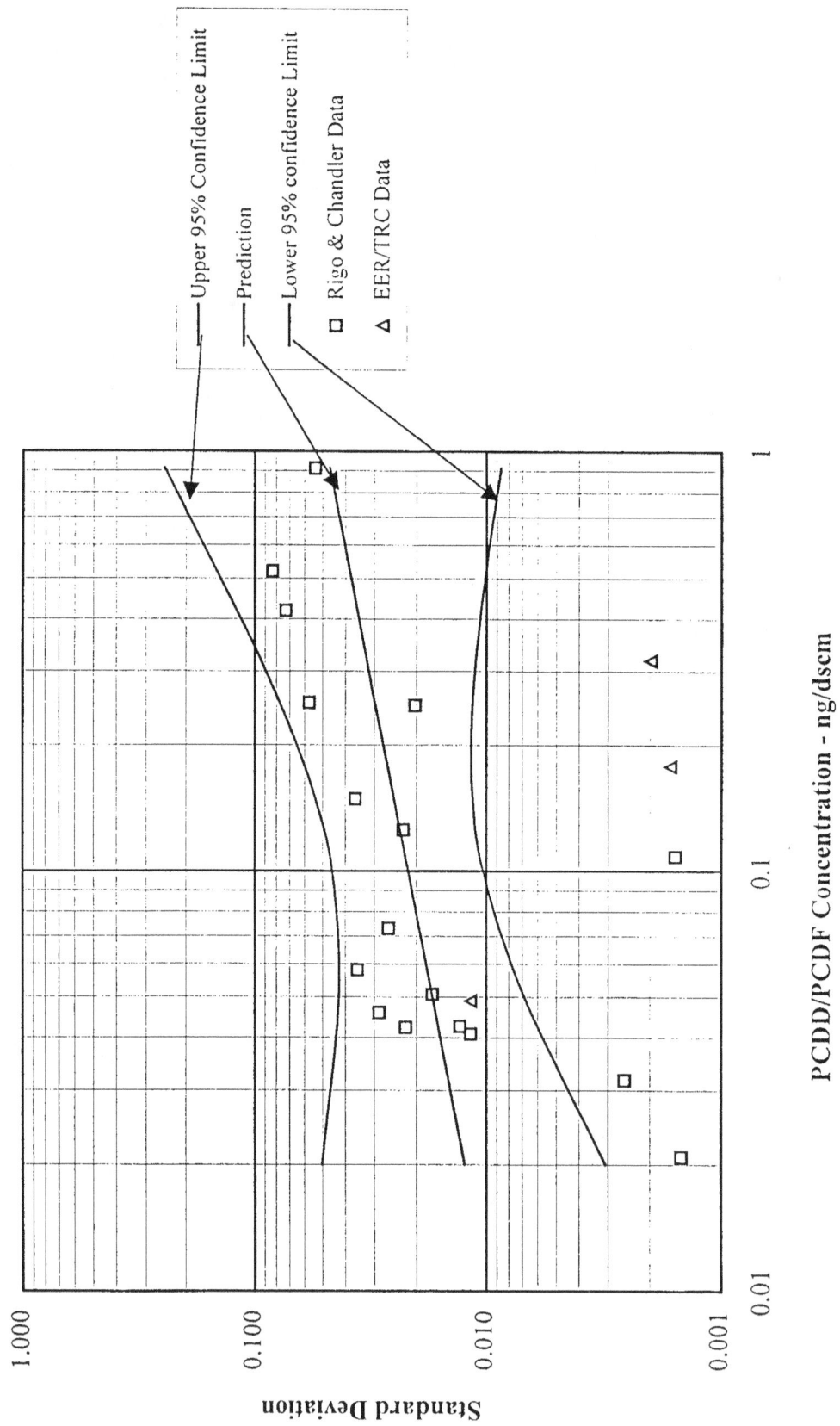

possible to construct figures illustrating the potential data scatter of Method 23 ITEQ measurements due to method imprecision. It is concluded, however, based on the t-statistic of 1.02 discussed above, that such figures would not provide a fair and balanced assessment of the Method.

Since the above analysis failed to determine a relationship between σ and C, the alternative approach is to reevaluate the data assuming that σ is a constant. The pooled analysis procedure is to first determine the pooled variance. The 22 individual values of S from Table 13 are squared, summed, and then divided by 22 to determine the pooled variance. The pooled variance = 0.000712. The square root of this parameter is taken to determine the pooled standard deviation; pooled S = 0.0267 with 22 degrees of freedom. Data from Table 1 in the Appendix provide factors for determining the 95% confidence bounds on σ. Those bounds are 0.0207 and 0.0374.

Figure 23 presents the pooled standard deviation and the 95% confidence intervals overlaid with the experimental data. Since the estimated standard deviation and the confidence intervals are constants, they are illustrated as straight lines in the Figure. The various precision metrics are presented in Figure 24. The data shown in this figure are normalized by the average concentration and thus show an inverse relationship with concentration.

If the characteristic standard deviation of Method 23 for ITEQ is equal to the pooled standard deviation, then measurement imprecision should cause 99 out of 100 future measurements to deviate from the true concentration no more than ±0.068 ng ITEQ/dscm. If the Method's characteristic standard deviation is more appropriately approximated by the upper 95% confidence bound, then method imprecision should cause 99 out of 100 future measurements to deviate from the true concentration by no more than ±0.095 ng ITEQ/dscm.

It is critically important that the above estimates for Method 23 imprecision be placed in perspective. Recent EPA regulations governing hazardous waste combustion systems and fossil fuel fired cement plants have set dioxin and furan emission limits of 0.2 ng ITEQ/dscm @ 7%

Figure 23. Pooled Regression Line and 95% Confidence Interval for
EPA Method 23 - PCDD/PCDF as ITEQ

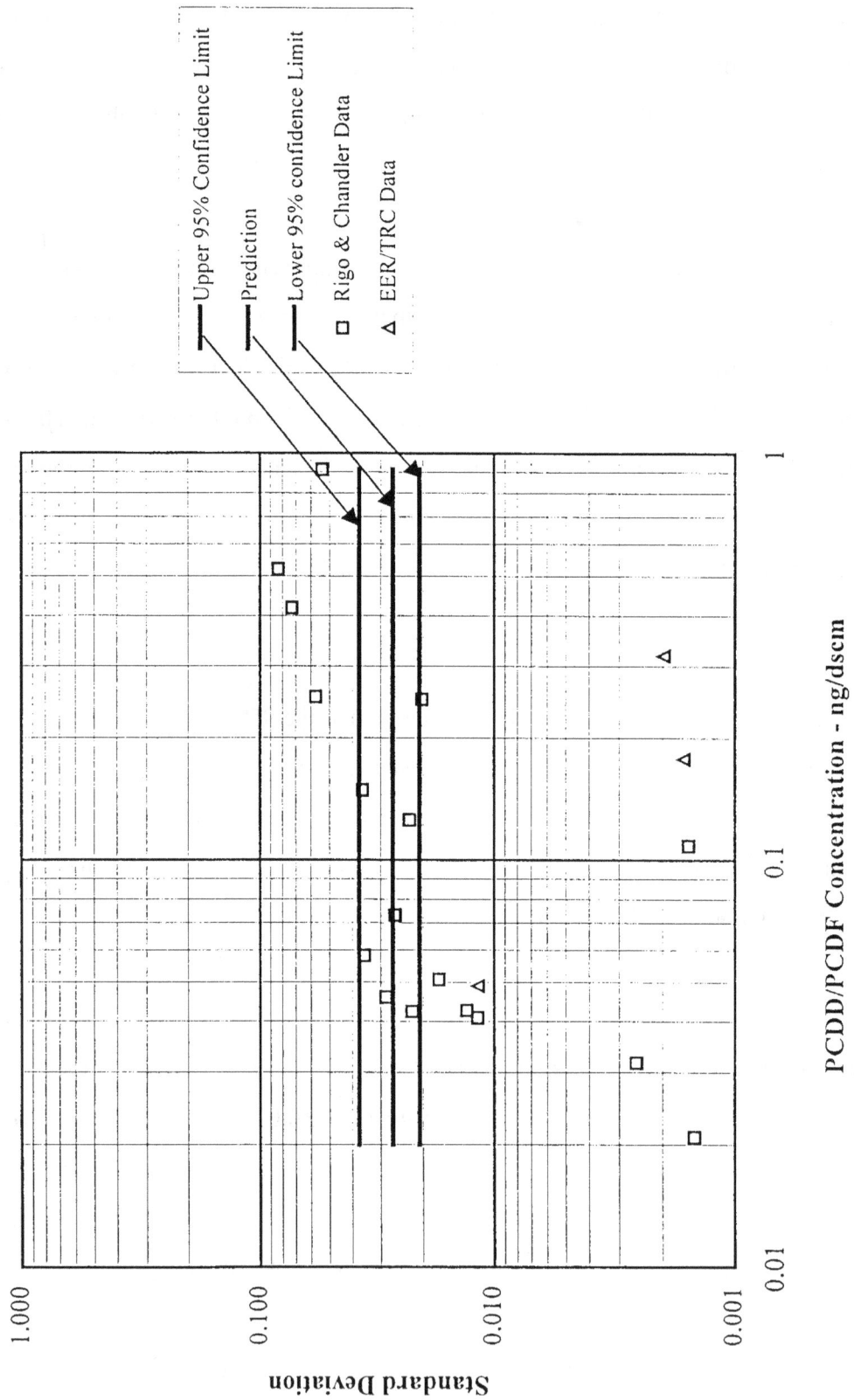

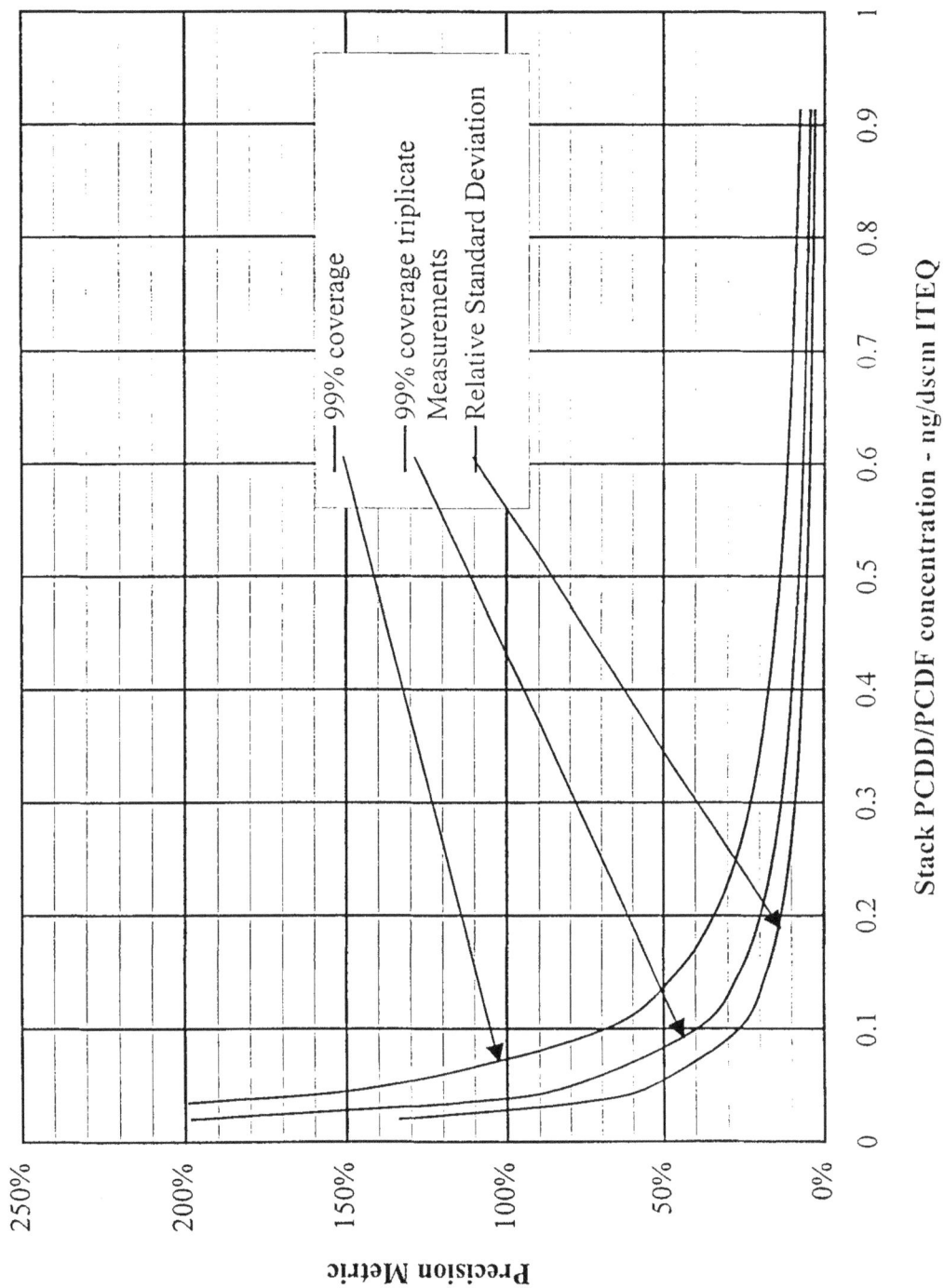

Figure 24. Method 23 Precision Based on Pooled Analysis of PCDD/PCDF Data as ITEQ

O_2. The potential range for future measurements from measurement imprecision is large relative to the standard. At the upper 95 % confidence limit, the possible range of a single measurement (minus 0.095 to plus 0.095 = 0.19) is essentially equal the standard itself.

This Page Intentionally Left Blank

This Page Intentionally Left Blank

6.0 EPA Method 26 for Hydrochloric Acid

The method for measurement of hydrochloric acid and chlorine gas is designated EPA Method 26 and is fully described in 40CFR Part 60 – Appendix A under the heading for the Method. The hardware arrangement for Method 26 is illustrated in Figure 25. Impingers in the back half of the train are filled with sulfuric acid to collect HCl while the sodium hydroxide impingers collect chlorine gas.

Multi-train data for Method 26 are available from three sources, summarized in Table 14. As indicated in the table, relatively high HCl concentration data (about 80 to 220 mg/dscm) are provided by the tests of Rigo and Chandler at an MWC facility in Pittsfield, MA (Rigo and Chandler, 1997). These data were collected using a quad train. Entropy Corp collected data, at intermediate HCl concentration (4 to 74 mg/dscm), as part of the EPA/OAQPS effort to validate Method 26 (Steinberger and Margeson, 1989). Two of the EPA/OAQPS tests (run number 11 and 12) were performed using four simultaneous sampling trains. Two of the four runs used midget impingers in what is now considered the standard Method 26 procedure. The other two trains used full size impingers typically associated with Method 5 trains. A significant bias was detected in the results from run number 12 and those results are not included in the ReMAP analysis. The remainder of the EPA/OAQPS test was conducted using dual trains with midget impingers. Finally, very low HCl concentration data (0.3 to 2.0 mg/dscm HCl) were collected by EER as part of an effort for EPA/OSW (EER, 1997). These EPA/OSW tests were executed using quad-trains. Data outlier analysis was performed on these data and all data points passed the outlier criteria set by the SPC procedures.

Figures 26 and 27 provide scatter plots of the available HCl data illustrating the standard deviation as a function of average HCl concentration. Figure 26 presents all of the available data while Figure 27 includes only data from the EPA/OAQPS and EPA/OSW tests (low HCl concentration data). Figure 28 presents these data as relative standard deviation. Data presented in all three of these figures have been corrected for small sample bias. As shown, when the

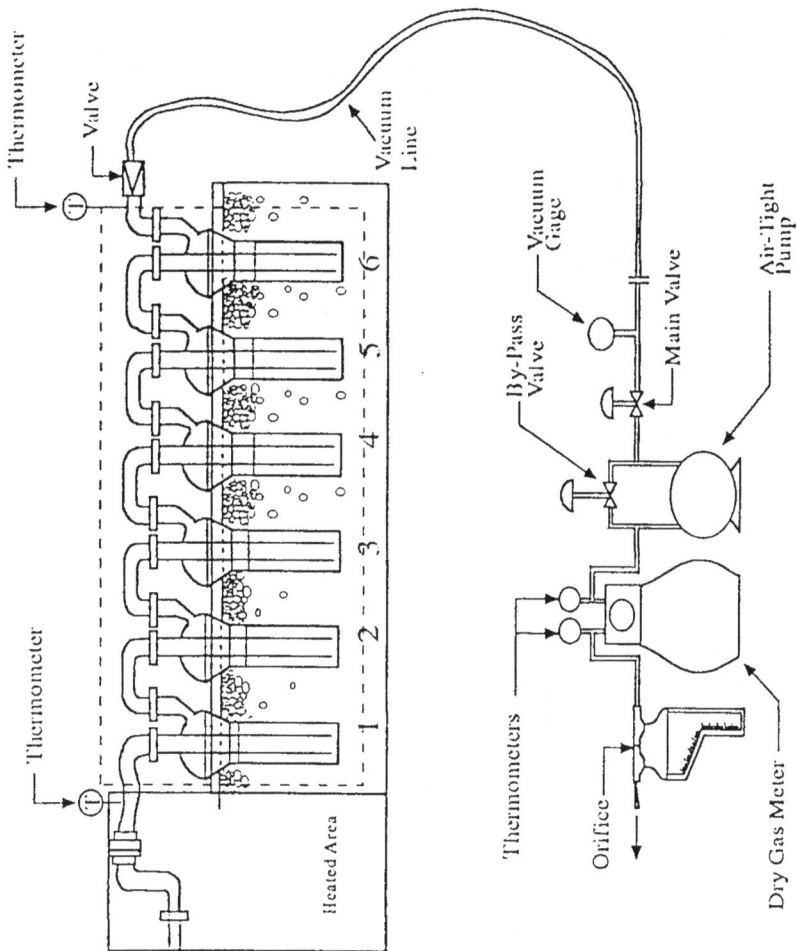

1. Modified G/S: 50ml 0.1 N H_2SO_4
2. Standard: 100ml 0.1 N H_2SO_4
3. Standard: 100ml 0.1 N H_2SO_4
4. Modified G/S: 100 ml 0.1 N NaOH
5. Modified G/S: 100 ml 0.1 N NaOH
6. Modified G/S: Silica Gel

Figure 25. Schematic of Method 26 sampling train.

Table 14. Method 26 Multi-Train Data and Standard Deviation for HCl

Data Source	Run Number	Train A	Train B	Train C	Train D	Average Concentration	Standard Deviation	S - Bias Corrected	RSD
Rigo & Chandler	1	128.92	140.45	147.26	153.04	142.42	10.367	11.248	7.90%
	2	210.80	215.42	221.16	221.78	217.29	5.187	5.628	2.59%
	3	161.46	175.73	152.64	149.51	159.83	11.742	12.740	7.97%
	4	72.54	76.24	84.85	85.56	79.80	6.430	6.976	8.74%
EPA-ORD Entropy	11	21.60	20.80	22.60	19.80	21.20	1.189	1.290	6.08%
	13	3.90	4.30			4.10	0.283	0.354	8.64%
	14	2.90	3.10			3.00	0.141	0.177	5.91%
	15	5.30	5.50			5.40	0.141	0.177	3.28%
	16	3.20	3.70			3.45	0.354	0.443	12.84%
	17	2.70	2.50			2.60	0.141	0.177	6.82%
	18	3.40	3.20			3.30	0.141	0.177	5.37%
	19	2.70	3.00			2.85	0.212	0.266	9.33%
	20	3.40	3.10			3.25	0.212	0.266	8.18%
	21	6.30	6.80			6.55	0.354	0.443	6.76%
	22	4.10	3.70			3.90	0.283	0.354	9.09%
	23	47.20	48.70			47.95	1.061	1.329	2.77%
	24	9.40	9.50			9.45	0.071	0.089	0.94%
	25	10.40	10.60			10.50	0.141	0.177	1.69%
	26	20.40	19.60			20.00	0.566	0.709	3.54%
	27	10.80	10.50			10.65	0.212	0.266	2.50%
	28	5.20	5.70			5.45	0.354	0.443	8.13%
	29	3.40	3.20			3.30	0.141	0.177	5.37%
EPA-OSW	10	1.47	1.36	1.38	1.07	1.32	0.176	0.191	14.52%
	11	0.57	0.47	0.62	0.46	0.53	0.079	0.086	16.34%
	12	0.43	0.35	0.51	0.58	0.47	0.100	0.108	23.23%
	13	0.43	0.48	0.57	0.43	0.48	0.068	0.074	15.45%
	14	0.68	0.60	0.47	0.38	0.53	0.135	0.147	27.60%
	15	0.33	0.27	0.28	0.31	0.30	0.031	0.034	11.43%
	16	0.28	0.17	0.18	0.21	0.21	0.050	0.054	26.00%
	17	0.20	0.18	0.19	0.20	0.19	0.011	0.012	6.42%
	18	0.22	0.17	0.24	0.26	0.22	0.037	0.040	17.70%

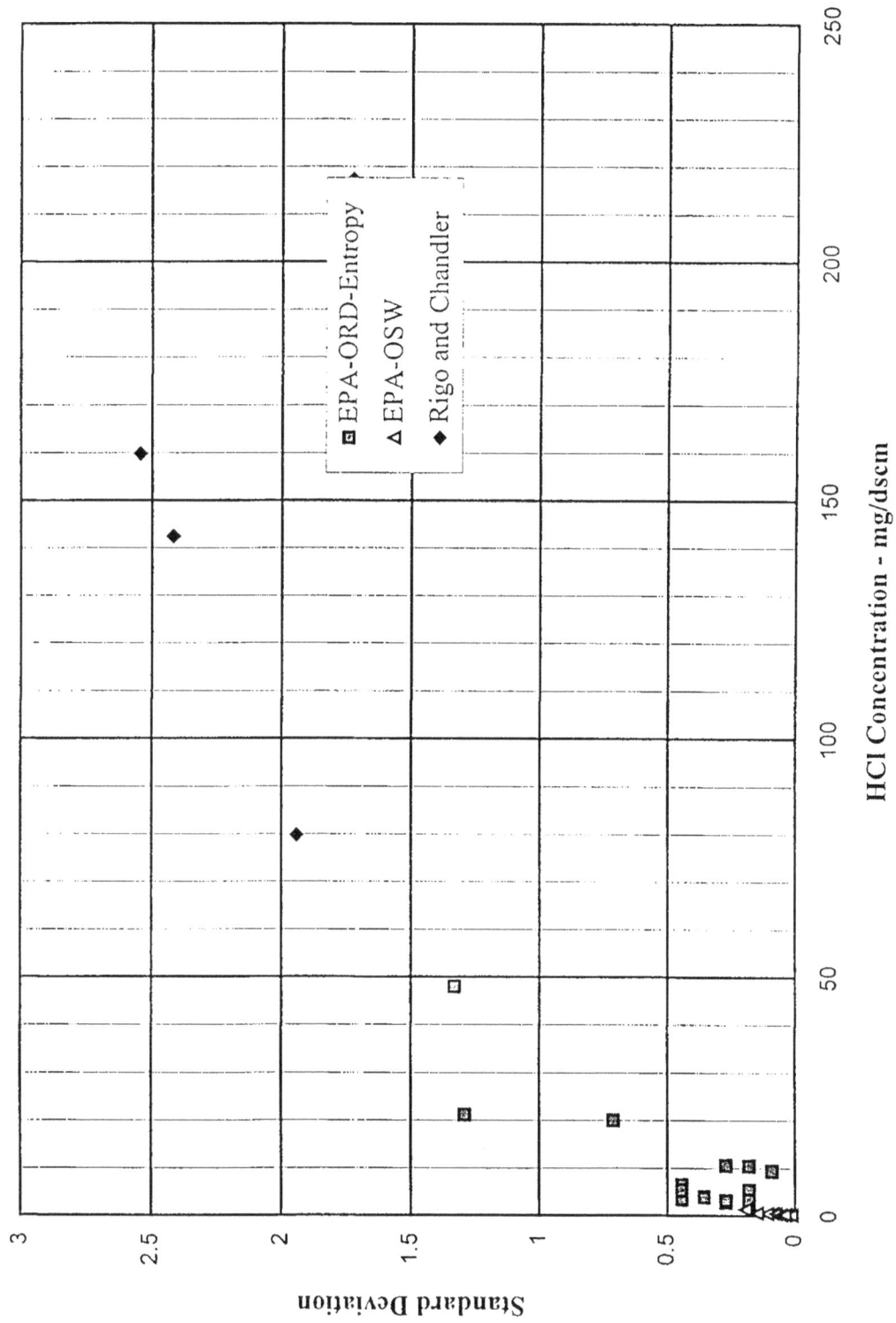

Figure 26. EPA Method 26 Data - HCl
Standard Deviation

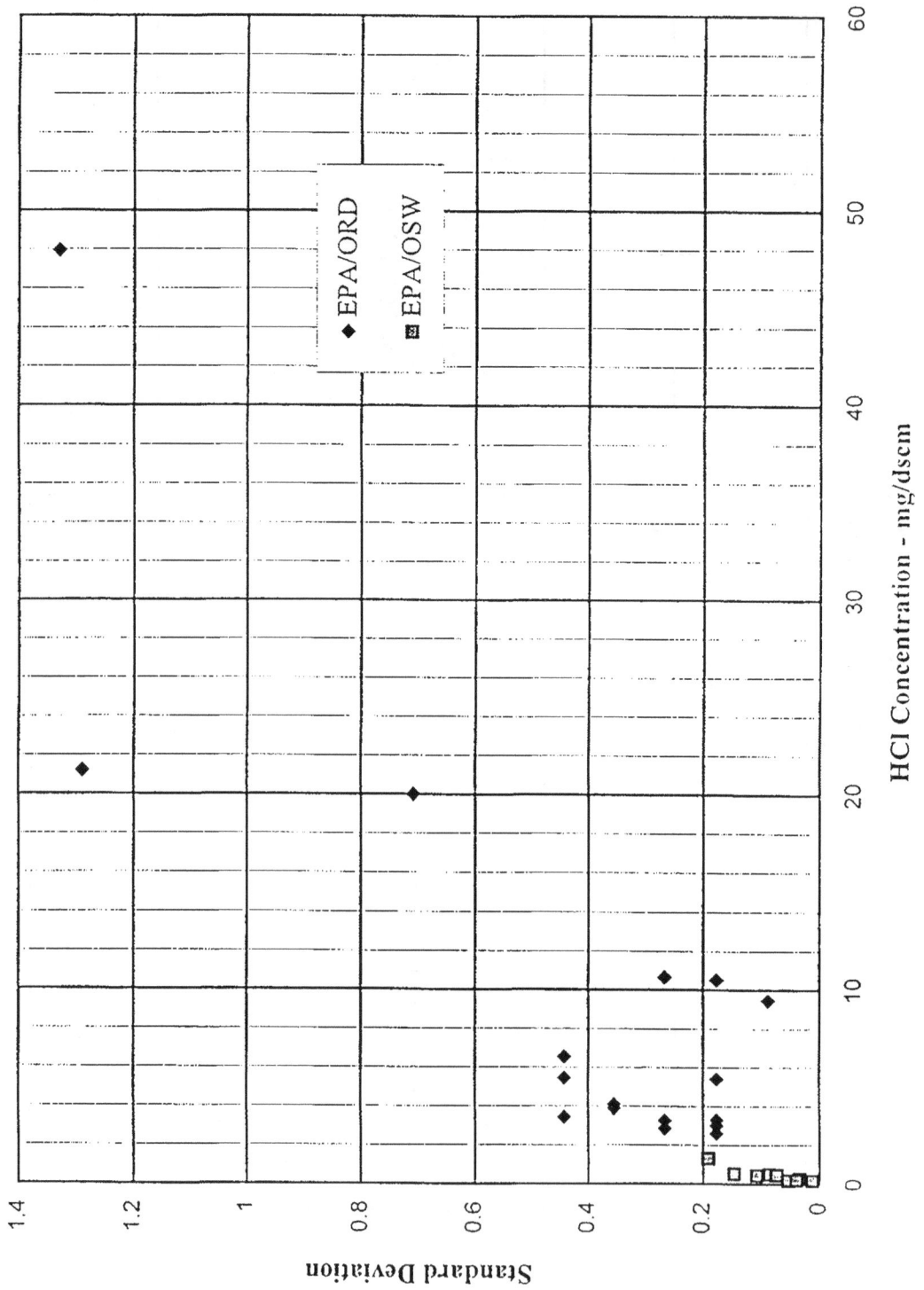

Figure 27. EPA Method 26 Data- Low and Intermediate HCl Concentration Standard Deviation

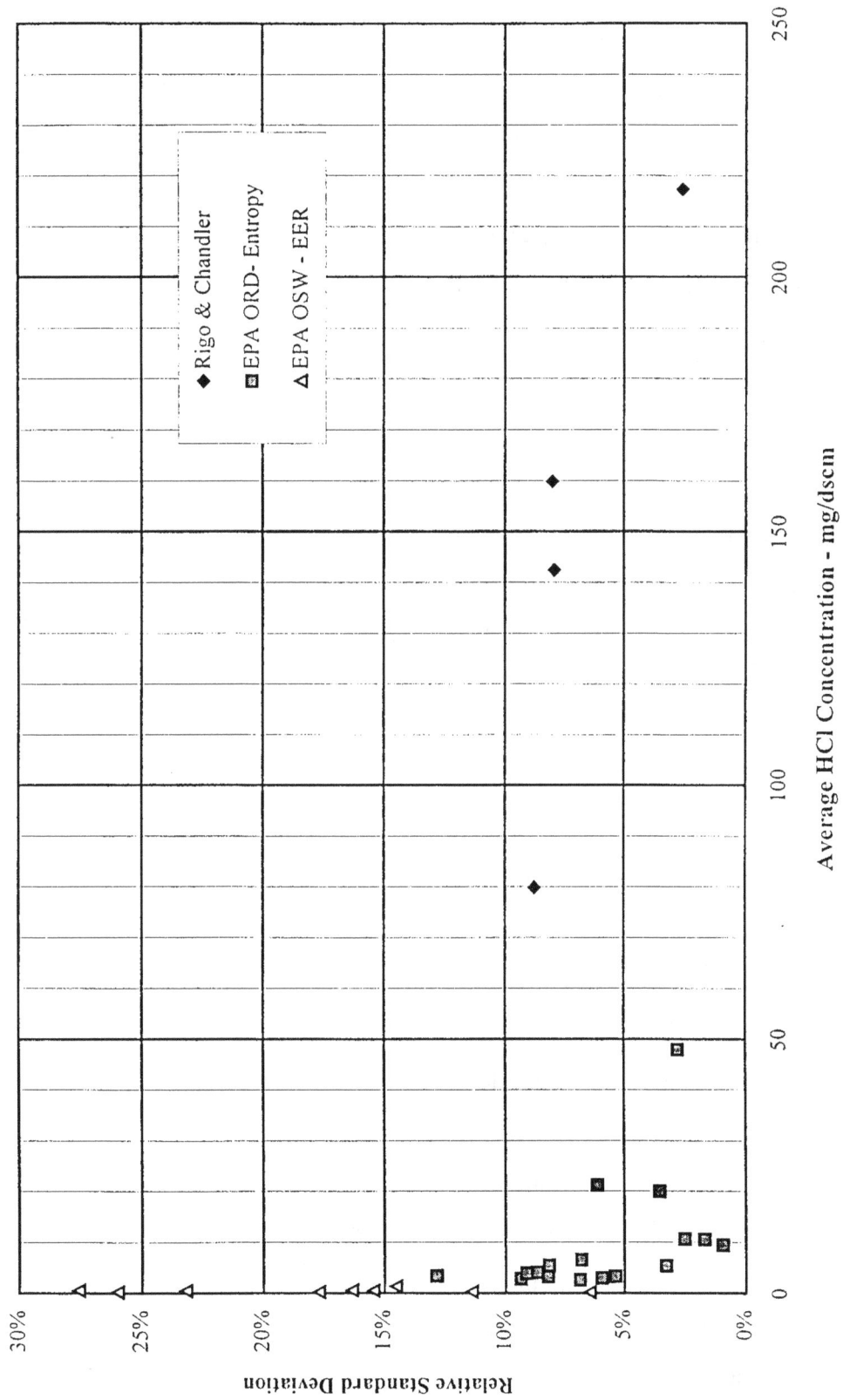

Figure 28. EPA Method 26 Data - HCl
Relative Standard Deviation

average HCl concentration is above about 10 mg/dscm, relative standard deviation (RSD) for the various runs is consistently below 10%. Below 10 mg/dscm, the RSD data tend to increase sharply. Figures 26 and 27 show that the sharp increase in RSD is caused by the rapidly decreasing value of the concentration rather than a sharp increase in standard deviation.

After applying appropriate weighting factors, the data in Table 14 were submitted to a weighted regression analysis and results indicate that the estimated standard deviation for the Method varies with HCl concentration according to the relation:

$$S \text{ (Method 26 for HCl)} = 0.15239 * C^{0.803}.$$

This equation includes both the small sample bias correction and the bias correction associated with the log-log transformation. The concentration term in this equation is in units of mg/dscm. The t-statistic for the regression is 17.22, which is well above the critical t-statistic for 95% confidence and 29 degrees of freedom (2.042). At the 95% confidence level, the value of the power coefficient is

$$P = 0.803 \pm 0.095 \text{ or between the limits of } 0.707 \text{ and } 0.898.$$

Figure 29 presents a plot of the data, along with the regression line and the upper and lower confidence limits. Note that the confidence intervals do not deviate significantly from the regression line. Using the regression equation to describe the variation of Est. σ with concentration, estimates can be developed for the probable variation in measurements associated with imprecision in the Method itself. Figure 30 presents results of those calculations including the variation in relative standard deviation, the estimated spread for 99 out of 100 single measurements, and the estimated spread for 99 out of 100 triplicate measurements. The anticipated spread for measurements is projected to be relatively close to the true stack concentration. For example, if Method 26 is applied to a stack containing 40 mg/dscm HCl, results from 99 out of 100 triplicate measurements are expected fall within 10.9 % of the true concentration. At lower concentrations, the range of future measurements is predicted to

Figure 29. Regression Line and 95% Confidence Intervals for
Method 26 HCl

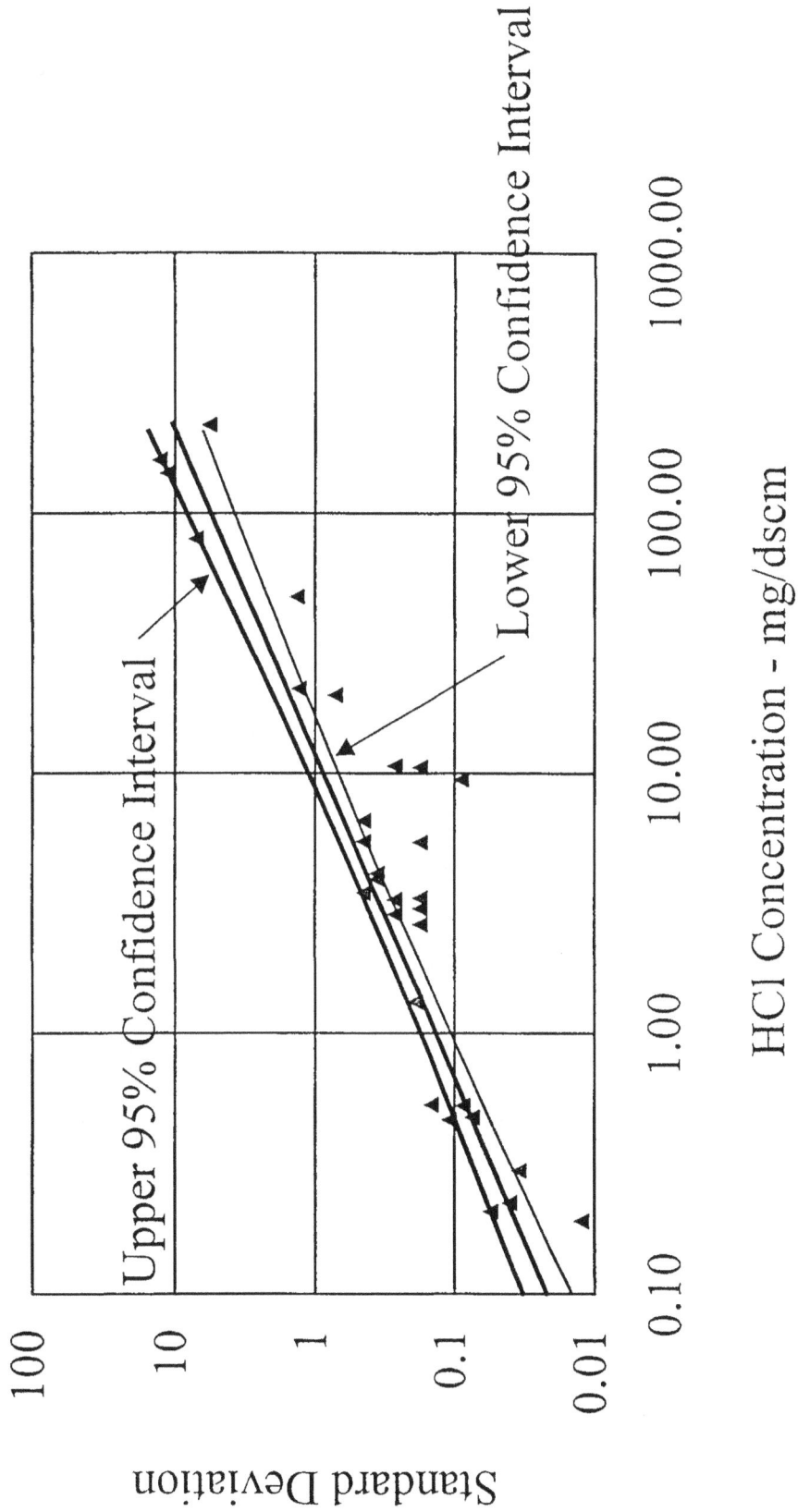

Upper 95% Confidence Interval

Lower 95% Confidence Interval

Standard Deviation

HCl Concentration - mg/dscm

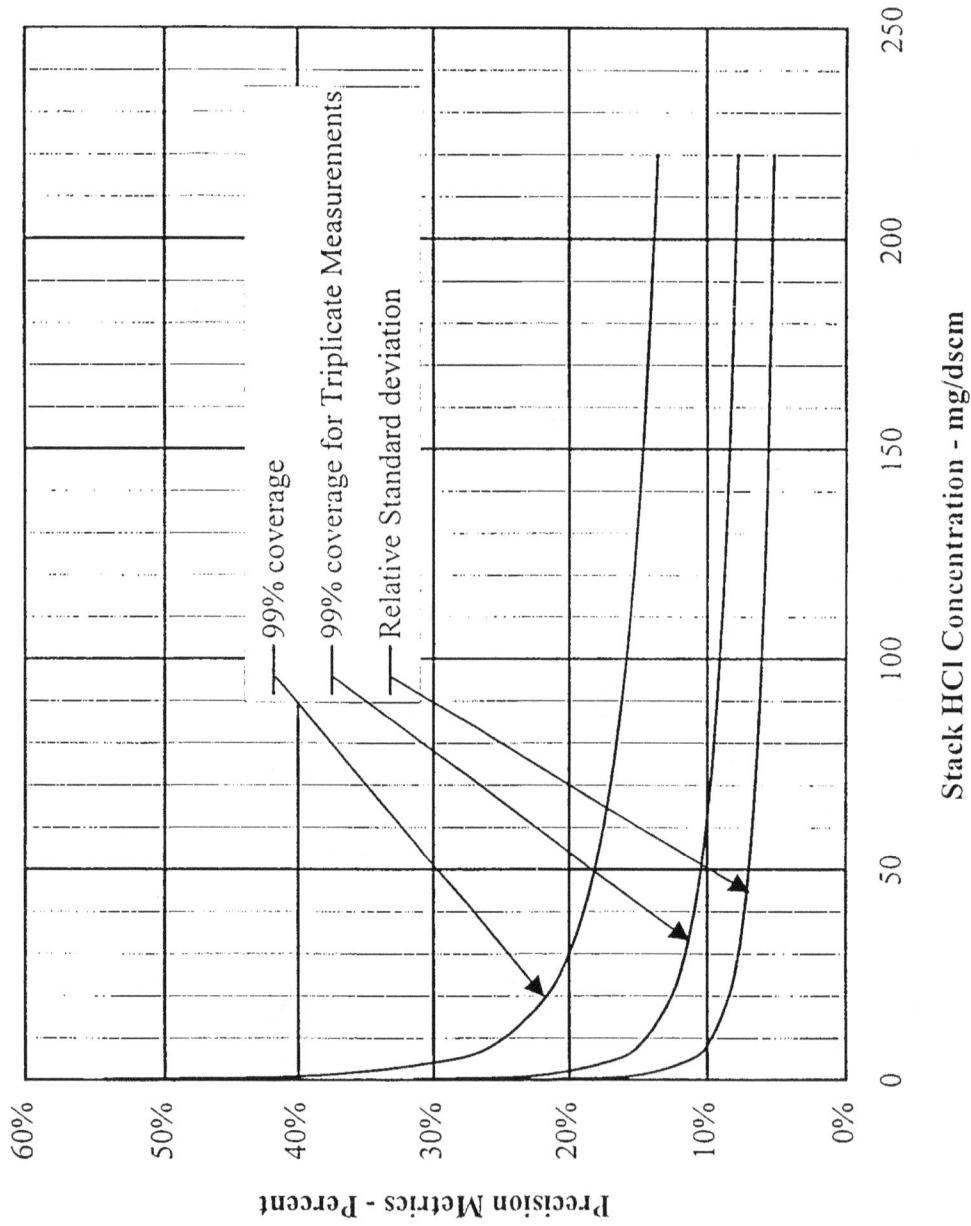

Figure 30. Precision Metrics for EPA Method 26 - HCl.

99

increase (as a percentage of the true concentration). However, even under the low concentration conditions, the actual spread in data from method imprecision is predicted to be relatively small. For example, 99 out of 100 triplicate measurements in a stack containing 1.0 mg/dscm HCl are expected to fall within the bounds of 0.80 and 1.20 mg/dscm.

The data in Figure 30 assumed that the standard deviation for Method 26 varied according to the regression equation. Figure 29 presented the 95% confidence bounds for the regression equation. Those data have been used to estimate the anticipated range for 99 out of 100 future measurements assuming the standard deviation varies with concentration according to the upper confidence interval, the regression equation, and according to the lower regression equation. Results for those calculations are presented in Figure 31.

The best estimate of the precision of Method 26 is provided by the regression analysis. Those data were presented graphically in Figures 30 and 31. To assist the reader, Table 15 presents these data in tabular form. The first column lists the true concentration of HCl in the stack (not corrected for excess oxygen). The second and third columns present the anticipated range for 99 out of 100 future single measurements (at the given true stack concentration). The fourth and fifth columns provide the anticipated range for 99 out of 100 future triplicate measurements.

Figure 31. Precision Estimates for HCl Measurements Using EPA Method 26

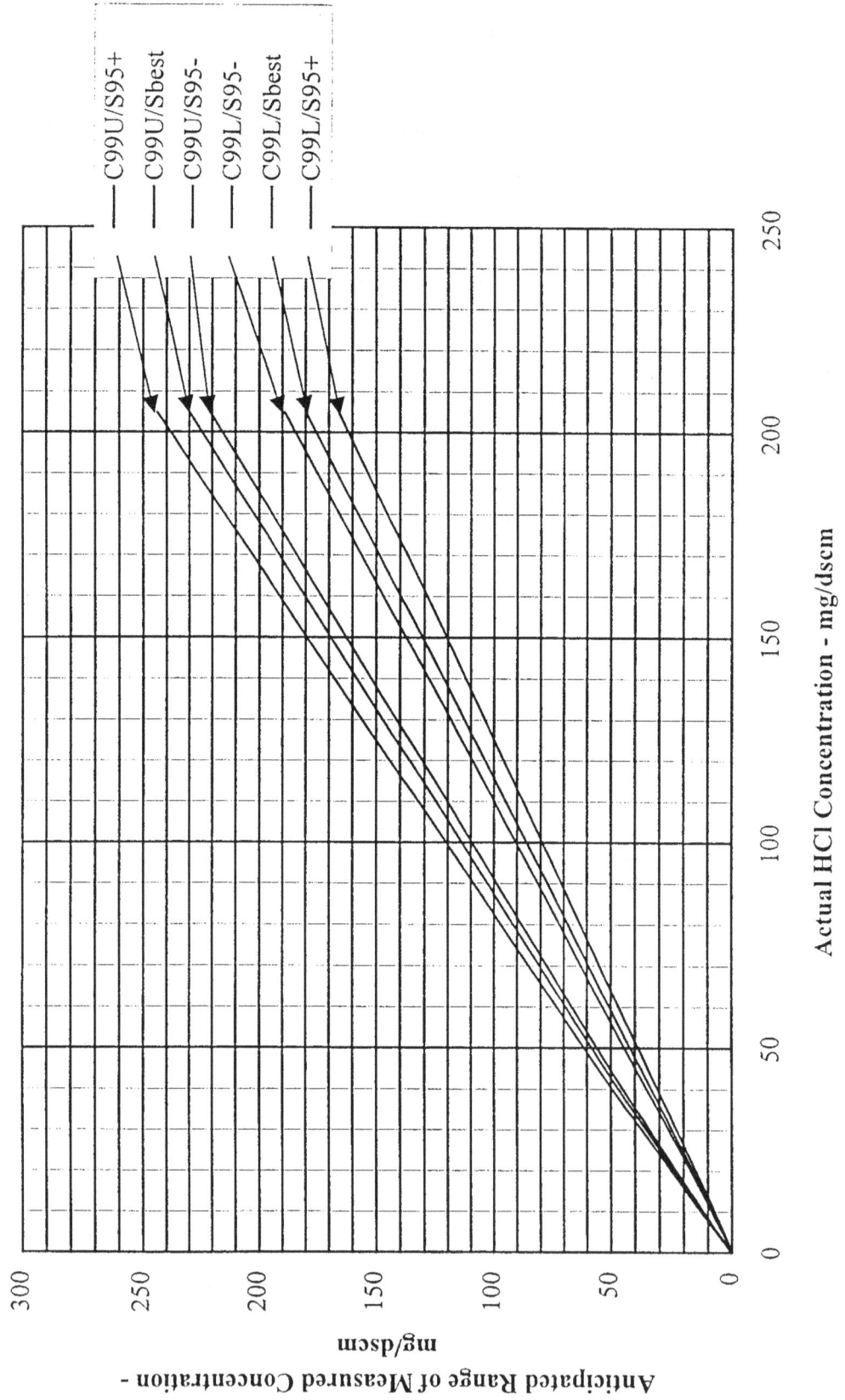

Legend:
- C99U/S95+
- C99U/Sbest
- C99U/S95-
- C99L/S95-
- C99L/Sbest
- C99L/S95+

X-axis: Actual HCl Concentration - mg/dscm

Y-axis: Anticipated Range of Measured Concentration - mg/dscm

Table 15. Range of Anticipated Future HCl Data

True Stack HCl Concentration mg/dscm	99 out of 100 Single Measurements		99 out of 100 Triplicate Measurements	
	Lower limit	Upper Limit	Lower limit	Upper Limit
1	0.65	1.35	0.80	1.20
5	3.72	6.28	4.26	5.74
10	7.76	12.2	8.71	11.29
20	16.1	23.9	17.75	22.25
30	24.6	35.4	26.88	33.12
40	33.2	46.8	36.07	43.93
50	41.9	58.1	45.30	54.70
60	50.6	69.4	54.56	65.44
70	59.3	80.7	63.85	76.15
80	68.1	91.9	73.15	86.85
90	77.0	103.0	82.47	97.53
100	85.8	114.2	91.81	108.19

7.0   EPA Methods 29, 101a and 101b for Mercury

Considerable effort has been expended developing methods for determination of mercury emissions from combustion sources. For details on the Methods, refer to 40 CFR Part 60 – Appendix A under the headings Method 29 and Method 101. Mercury is emitted from combustion sources as either the base metal or in the +2 valence-state. It may be in the vapor-state or it may be associated with solid phase material. At least three EPA Methods have been published or proposed and one additional method is under active development. Figure 32 illustrates the hardware used for Method 101a, which is the simplest of the mercury procedures. The sample gas is passed through a heated filter and then goes to impingers filed with a $KMnO4-10\%H_2SO_4$ (permanganate) solution. After the sampling event the filter and permanganate solutions are digested and a single combined measurement of total mercury concentration is generated. Method 29 was developed as a multi-metal measurement method, including mercury. The hardware configuration for Method 29, illustrated in Figure 33, includes a pair of nitric acid impingers followed by permanganate impingers. The various fractions are analyzed separately for a wide range of metals and results summed to develop a reported stack concentration. Finally, Method 101b, illustrated in Figure 34 is a simple variant on Method 29. Method 101b replaces one of the nitric acid impingers in Method 29 with two water impingers. The various fractions are analyzed separately in an attempt to determine the split between Hg^0 and Hg^{+2}. The total mercury concentration is determined by adding the catch from the various fractions.

An extensive array of multi-train data is available describing the precision of these three mercury measurement methods. The first set of results was developed by EPA/OAQPS as part of the validation of Method 29 (Radian, 1992). Those tests, conducted by Radian Corp., were performed at a municipal waste combustor operated by American Ref-Fuel in New Jersey. That data set includes 8 quad-train runs with stack mercury concentration ranging from 130 to 575 µg/dscm. The next set of data comes from the EPA efforts to validate Method 101b (EER, 1997). Extensive testing was conducted at a hazardous waste burning cement kiln. Testing consisted of triplicate train runs using both Methods 101b and Method 29 (test series 1) and quad

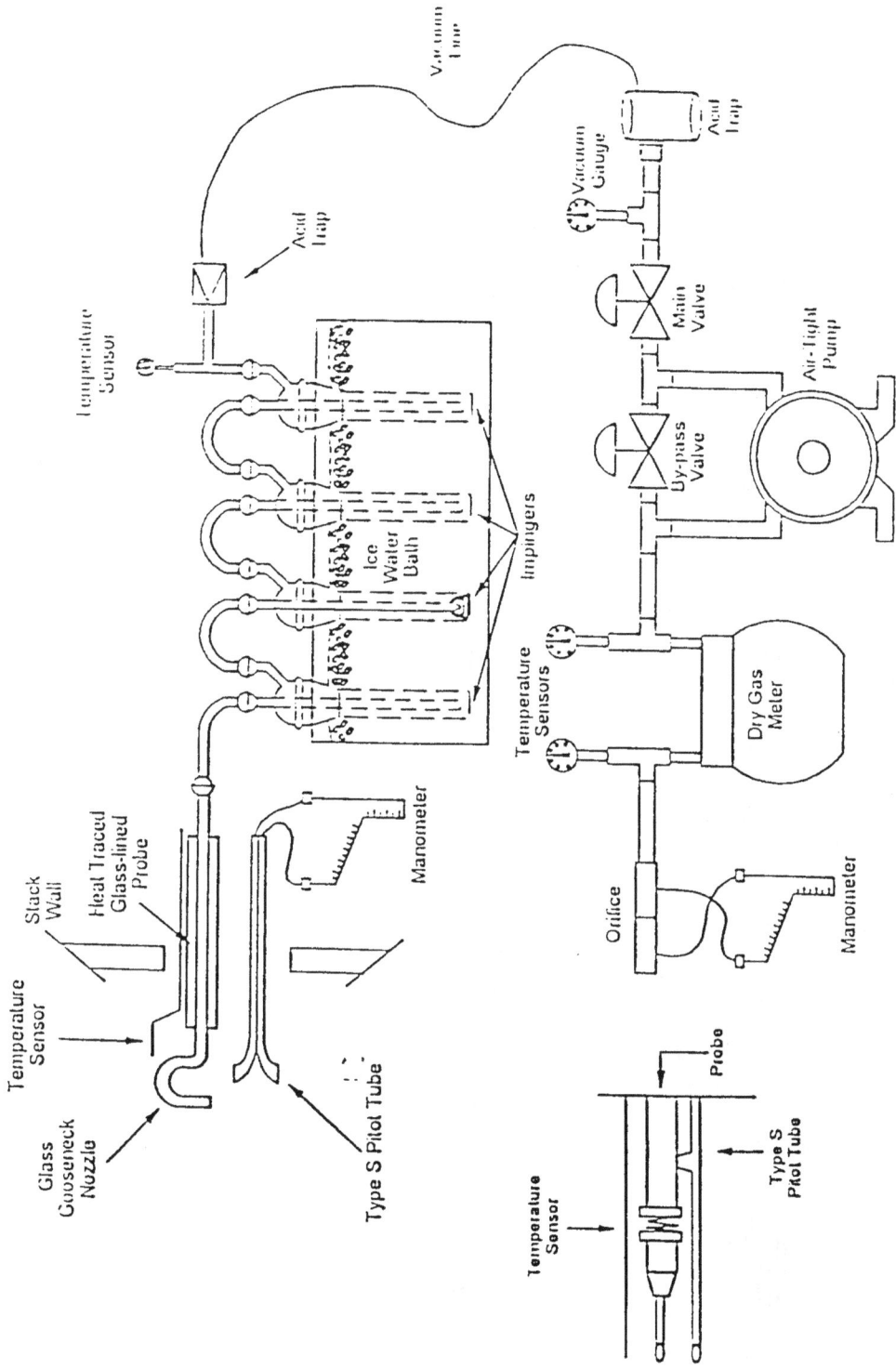

Figure 32. Schematic of Method 101A sampling train.

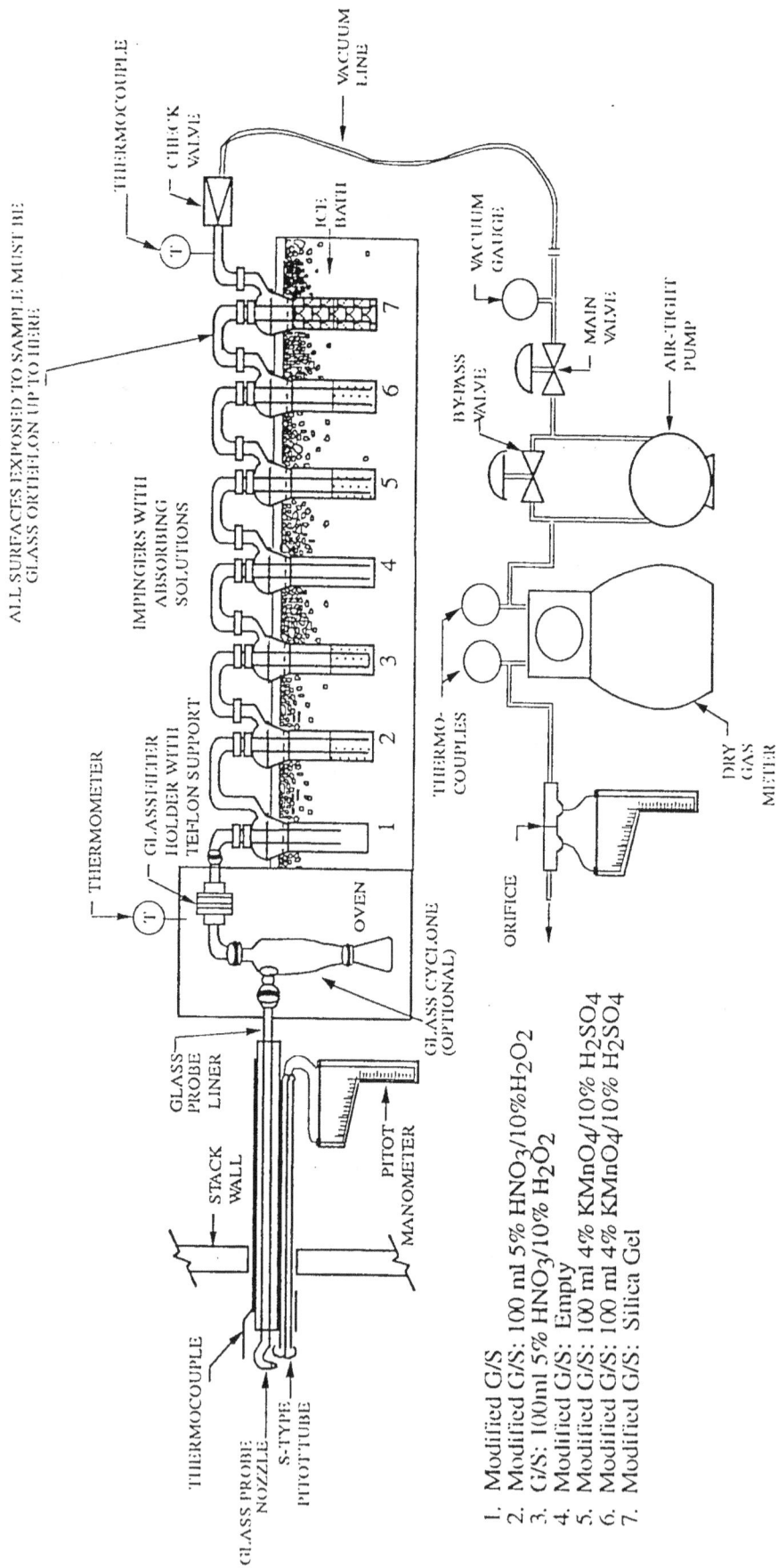

ALL SURFACES EXPOSED TO SAMPLE MUST BE
GLASS OR TEFLON UP TO HERE

THERMOCOUPLE

CHECK VALVE

VACUUM LINE

THERMOMETER

ICE BATH

IMPINGERS WITH ABSORBING SOLUTIONS

GLASS FILTER HOLDER WITH TEFLON SUPPORT

VACUUM GAUGE

BY-PASS VALVE

MAIN VALVE

AIR-TIGHT PUMP

THERMO-COUPLES

DRY GAS METER

OVEN

GLASS CYCLONE (OPTIONAL)

ORIFICE

THERMOMETER

GLASS PROBE LINER

STACK WALL

PITOT MANOMETER

THERMOCOUPLE

GLASS PROBE NOZZLE

S-TYPE PITOT TUBE

1. Modified G/S
2. Modified G/S: 100 ml 5% HNO_3/10%H_2O_2
3. G/S: 100ml 5% HNO_3/10% H_2O_2
4. Modified G/S: Empty
5. Modified G/S: 100 ml 4% $KMnO_4$/10% H_2SO_4
6. Modified G/S: 100 ml 4% $KMnO_4$/10% H_2SO_4
7. Modified G/S: Silica Gel

Figure 33. Schematic of Method 29 sampling train.

105

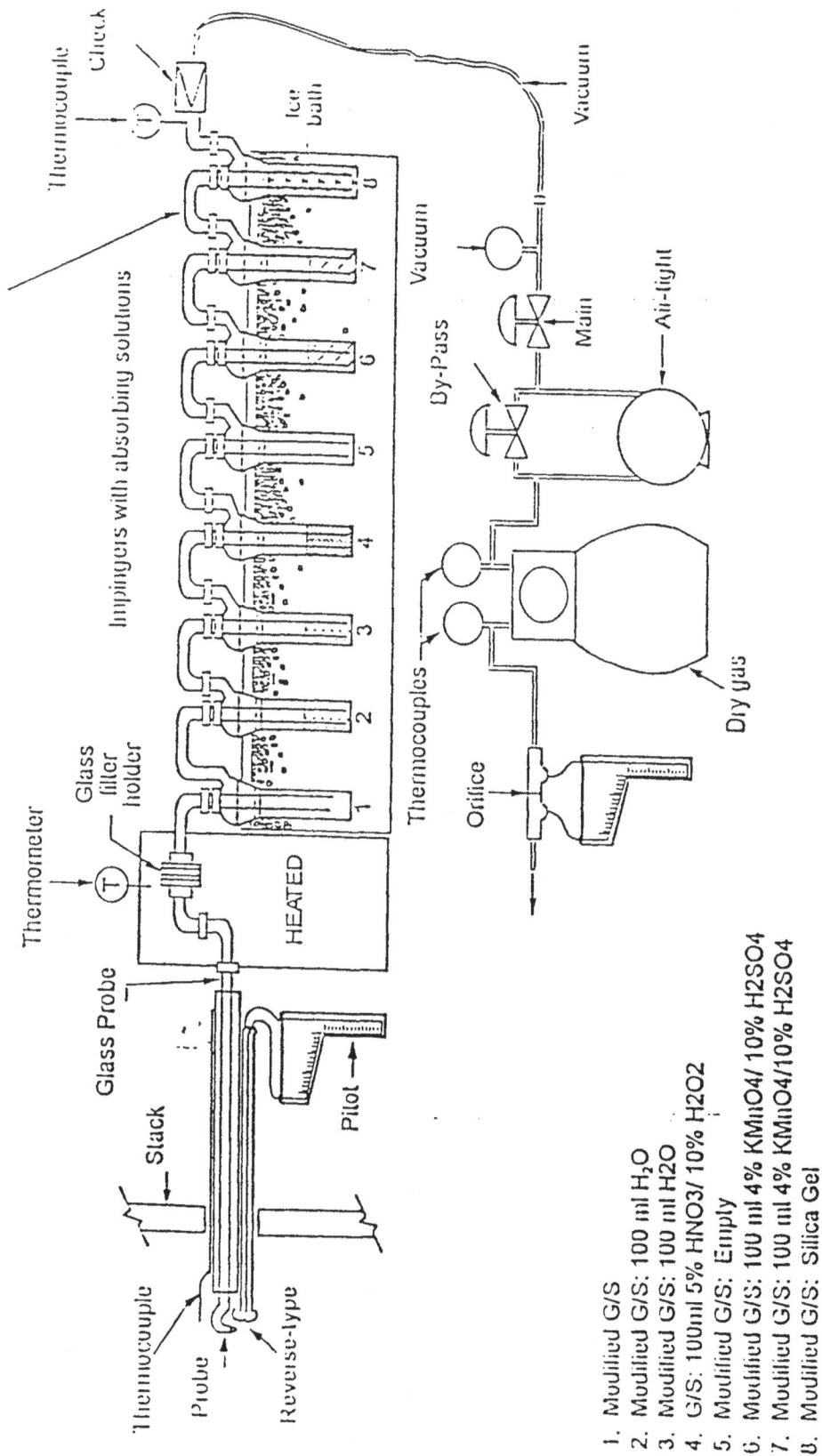

Figure 34. Schematic of Method 101B sampling train.

All glass sample exposed surface to here.

Thermocouple
Check
Thermocouple
Ice bath
Vacuum

Impingers with absorbing solutions

Vacuum

Dy-Pass

Main

Air-tight

Thermocouples

Orifice

Dry gas

Glass filter holder

Thermometer

HEATED

Glass Probe

Stack

Thermocouple
Probe
Reverse-type

Pilot

1. Modified G/S
2. Modified G/S: 100 ml H_2O
3. Modified G/S: 100 ml $H2O$
4. G/S: 100ml 5% HNO3/ 10% $H2O2$
5. Modified G/S: Empty
6. Modified G/S: 100 ml 4% KMnO4/ 10% $H2SO4$
7. Modified G/S: 100 ml 4% KMnO4/10% $H2SO4$
8. Modified G/S: Silica Gel

train runs using only Method 101b (test series 2.). The triplicate train tests consisted of a Method 101b train located in the breaching to the stack (downstream of all air pollution controls) as well as Methods 101b and 29 trains located at the same plane of the stack. These three measurements were all treated as simultaneous but it is distinctly possible that the diverse locations could contribute to an increased apparent method imprecision for this data set. For the quad train tests two of the lines were dynamically spiked with Hg^0 and $HgCl_2$. For the ReMAP analysis, only the unspiked data from the quad trains have been used.

A third set of multi-train data was provided by the Rigo and Chandler tests on the MWC facility in Pittsfield, MA. These tests provide dual-train measurements using EPA Method 29. A fourth set of multi-train data is provided in EPA tests at the Stanislaus County MWC (Radian, 1992). In those tests, two Method 29 measurements were made simultaneously in the stack. From these tests, the only dual-train metal data available from the EPA were for mercury.

A fifth set of data is provided by tests performed at EPA's research facilities in RTP, NC (EPA, 1998). There are several key features to the data from these tests that should be pointed out as part of the ReMAP data analysis. First, numerous Method 29 tests were conducted as part of an effort to assess performance of multi-metal continuous emission monitors. The EPA combustion facility at RTP is a pilot-scale rotary kiln incinerator with a full RCRA permit. As part of the CEMS assessment, the metal content of the waste feed was varied to adjust the range of metal concentration in the stack. The physical arrangement of the stack causes the flow to travel from a mezzanine level, down one floor, and then horizontally to a final clean up baghouse. A series of multi-metals CEMS were installed in the vertical portion of this duct. Method 29 trains were located in the horizontal duct runs at the mezzanine level (before the CEMS) and at the floor level below. Manual method data were collected simultaneously. Because of the wide concentration range of these paired data, these results are extremely important to the overall ReMAP effort. A careful assessment of the data indicates that there is a distinct bias in the results. That bias is only detectable because of the large number of metals being measured. [Material presented in the Appendix discusses the data trends in some detail.] The essence of imprecision is that simultaneous measurements of the same stack gas yield different measurement results. What is unusual about the EPA pilot-scale tests is that the reported concentrations for all of the metals tend to move in concert. If the mezzanine train indicates a higher concentration for one metal (relative to the floor level sampling train), it will also indicate a higher concentration for the other metals as well. Both EPA and the ASME ReMAP team conducted a thorough assessment to determine the root cause of this bias. Unfortunately no firm conclusions were forth coming. It seems apparent that the source of the problem can be traced to the sample collection process or the hardware used to collect the sample as opposed to the laboratory analytical procedures. Other potential data or process effects were also investigated. Though there is a distinct bias, the data from these tests have been used as provided by the EPA researchers. The effect may be to slightly increase the indicated standard deviation of individual method 29 data points, but the wide range of data greatly improves the overall quality of the method precision estimates.

Tables 16a and 16b provide a summary of all the mercury emission data noted above. The first step in the analysis is to perform an outlier analysis of the data following the SPC procedures outlined previously. The data were grouped into sets with concentrations above and below 100μg/dscm. The analysis identified four data points as having abnormally large spans. Those runs included Run Number 9 from the EPA/OSW tests (second series) and Runs 5-2, 8-3 and 9-1 from the Stanislaus County tests. The span for Run 9 from the EPA/OSW second test series is clearly out of line with the remainder of data from that test series as well as with other mercury data with concentrations below 100 μg/dscm. Based on the values of RSD there is also reason to suspect Runs 6-3, 8-1, and 9-2 as potential outliers. However, the overall bias for the ReMAP program and generally accepted statistical analysis procedure is to retain all data unless there is a clear rationale for data elimination. Accordingly, only the four measurements identified as having abnormally large data spans have been eliminated from further analysis. Figures 35 and 36 present the data from Tables 16a and 16b showing scatter plots of standard deviation and relative standard deviation as a function of the run average concentration.

Data in Tables 16a and 16b (after eliminating outliers) have been subjected to regression analysis. Results of the analysis are presented in Figure 37 showing the regression line, the upper and lower

Table 16a. Methods 29 and 101 Data for Mercury

Data Source	Run Number	Train A	Train B	Train C	Train D	Average Concentration	Standard Deviation	S - Bias Corrected	RSD
EPA/ OAQPS	1	213.81	200.70	222.16	184.19	205.22	16.56	17.97	8.76%
	2	165.29	173.55	176.20	191.98	176.75	11.16	12.11	6.85%
	3	284.67	295.45	316.72	312.72	302.39	14.99	16.27	5.38%
	4	584.68	584.42	595.39	532.72	574.30	28.19	30.59	5.33%
	5	186.39	175.39	193.04	163.12	179.49	13.12	14.23	7.93%
	6	224.20	225.01	212.03	211.84	218.27	7.32	7.94	3.64%
	7	185.08	178.94	172.04	179.55	178.90	5.35	5.80	3.24%
	8	132.24	129.50	124.20	132.62	129.64	3.88	4.21	3.25%
EPA-OSW	1	26.56	20.27			23.41	4.45	5.57	23.80%
	2	27.55	20.71	22.61		23.62	3.53	3.98	16.86%
	3	23.84	21.66	21.07		22.19	1.46	1.64	7.40%
	4	26.51	24.18	22.31		24.33	2.11	2.38	9.76%
	5	26.42	25.75	21.23		24.47	2.82	3.19	13.02%
	6	24.57	23.43	23.00		23.67	0.81	0.91	3.87%
	7	25.56	23.24	21.73		23.51	1.93	2.18	9.26%
	8	20.96	18.65	18.38		19.33	1.42	1.60	8.28%
	9	25.98	19.12	21.48		22.19	3.49	3.93	17.72%
	2	13.41	14.20			13.81	0.56	0.70	5.07%
	3	23.64	19.84			21.74	2.69	3.37	15.49%
	4	27.26	24.36			25.81	2.05	2.57	9.96%
	5	19.69	15.70			17.70	2.82	3.54	19.98%
	6	25.52	29.35			27.44	2.71	3.39	12.37%
	7	32.88	28.27			30.58	3.26	4.08	13.36%
	8	26.39	31.93			29.16	3.92	4.91	16.83%
	9	44.64	22.59			33.62	15.59	19.54	58.12%
	10	24.41	22.30			23.36	1.49	1.87	8.00%

Table 16b. Methods 29 and 101 Data For Mercury

Data Source	Run Number	Train A	Train B	Average Concentration	Standard Deviation	S - Bias Corrected	RSD
Rigo & Chandler	1	60.63	53.13	56.88	5.30	6.64	11.68%
	3	8.39	9.99	9.19	1.13	1.42	15.41%
	6	3.33	6.43	4.88	2.19	2.75	56.26%
	7	9.63	7.07	8.35	1.81	2.27	27.19%
	8	136.28	136.75	136.52	0.33	0.41	0.30%
	9	108.62	110.16	109.39	1.09	1.36	1.24%
	10	73.35	77.87	75.61	3.20	4.00	5.30%
	11	20.41	24.04	22.22	2.57	3.21	14.46%
	12	20.57	20.54	20.55	0.02	0.02	0.11%
	13	55.45	51.99	53.72	2.45	3.07	5.71%
	14	12.70	13.73	13.22	0.73	0.91	6.90%
	15	13.99	14.78	14.38	0.56	0.70	4.87%
	16	10.13	10.20	10.17	0.05	0.06	0.62%
	17	14.40	14.63	14.51	0.17	0.21	1.43%
	18	12.24	12.14	12.19	0.07	0.08	0.68%
	19	8.46	9.33	8.89	0.62	0.77	8.68%
EPA Pilot Scale MM CEM Demo	1	184.76	176.08	180.42	6.14	7.69	4.26%
	2	179.04	174.14	176.59	3.47	4.35	2.46%
	3	174.79	226.08	200.44	36.27	45.45	22.67%
	5	169.07	134.95	152.01	24.13	30.23	19.89%
	1	168.65	180.16	174.40	8.13	10.19	5.84%
	2	194.60	194.19	194.39	0.29	0.36	0.19%
	4	206.92	194.09	200.50	9.07	11.37	5.67%
	5	205.52	206.33	205.93	0.57	0.72	0.35%
	1	37.32	38.03	37.67	0.50	0.63	1.66%
	2	40.80	36.29	38.54	3.19	3.99	10.36%
	3	35.68	25.13	30.41	7.46	9.34	30.72%
	4	28.58	29.80	29.19	0.86	1.08	3.71%
	5	37.61	52.96	45.29	10.85	13.60	30.03%
	1	32.77	39.20	35.98	4.55	5.70	15.83%
	2	40.22	41.09	40.66	0.62	0.77	1.90%
	3	36.14	36.70	36.42	0.39	0.49	1.36%
	4	41.84	45.47	43.66	2.57	3.22	7.37%
	5	39.27	41.91	40.59	1.87	2.34	5.77%
Stanislaus County MWC	4-1	483.00	405.00	444.00	55.15	69.11	15.56%
	5-1	807.00	759.00	783.00	33.94	42.53	5.43%
	5-2	466.50	706.50	586.50	169.71	212.64	36.26%
	5-3	736.50	637.50	687.00	70.00	87.71	12.77%
	6-2	72.00	70.50	71.25	1.06	1.33	1.87%
	6-3	25.50	57.00	41.25	22.27	27.91	67.66%
	6-4	97.50	78.00	87.75	13.79	17.28	19.69%
	8-1	198.00	96.00	147.00	72.12	90.37	61.48%
	8-2	138.00	111.00	124.50	19.09	23.92	19.21%
	8-3	57.00	118.50	87.75	43.49	54.49	62.10%
	9-1	198.00	342.00	270.00	101.82	127.58	47.25%
	9-2	121.50	214.50	168.00	65.76	82.40	49.05%
	9-3	150.00	196.50	173.25	32.88	41.20	23.78%

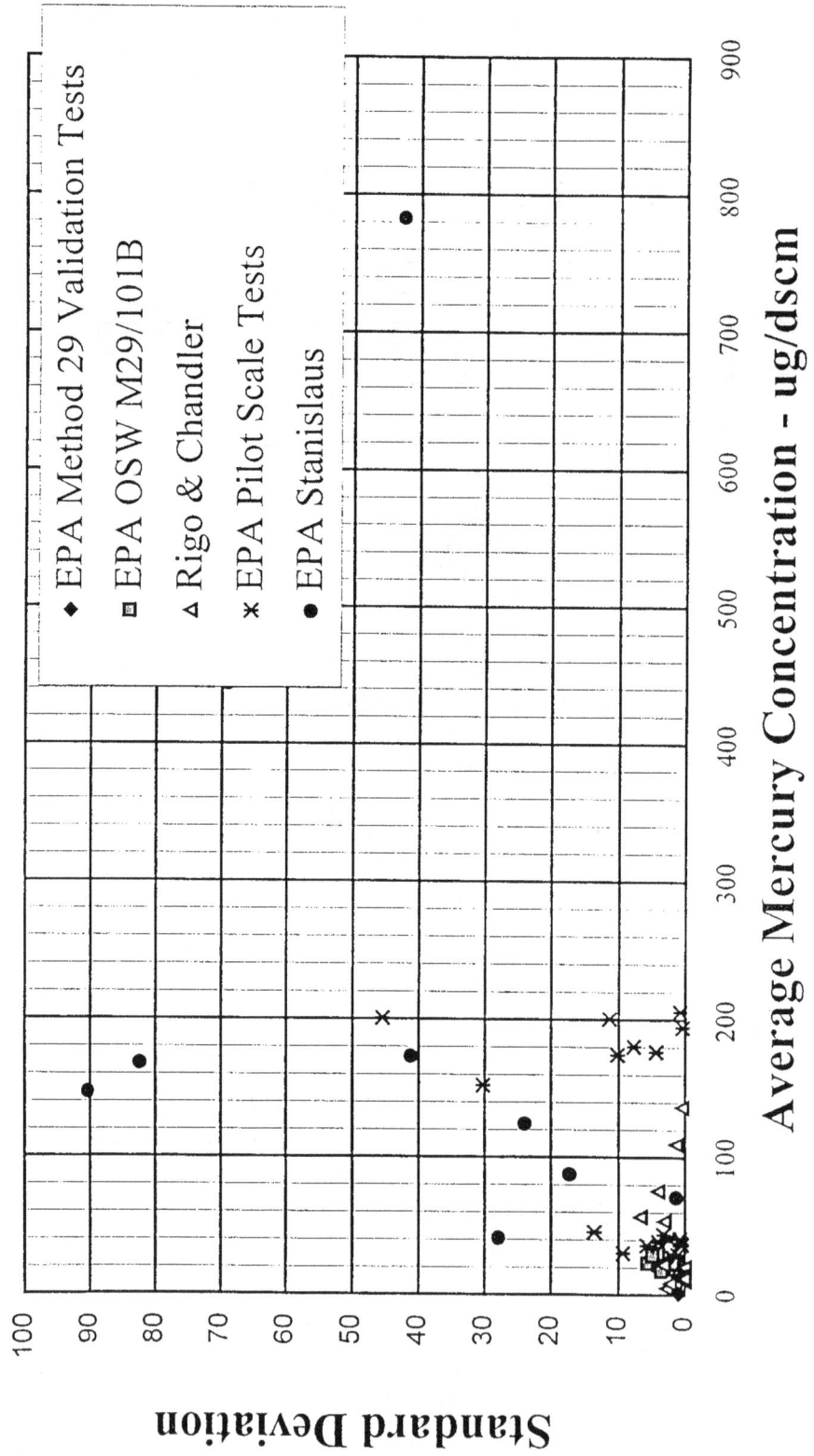

Figure 35. EPA Method 29/101a/101b Data – Total Mercury Standard Deviation

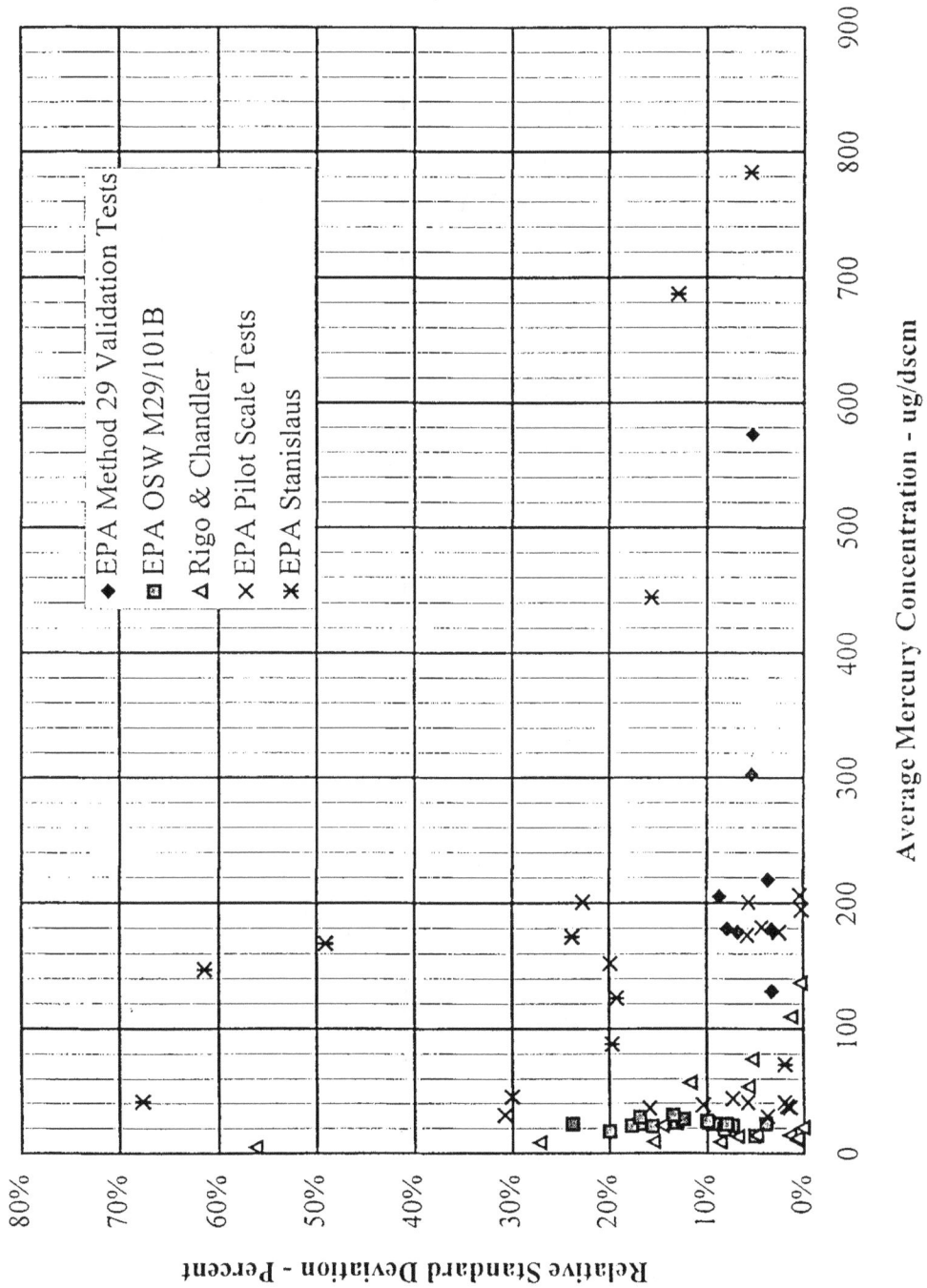

Figure 36. EPA Method 29/101a/101b Data - Total Mercury
Relative Standard Deviation

Figure 37. Regression Line and 95% Confidence Interval EPA Methods 29, 101a and 101b For Total Mercury

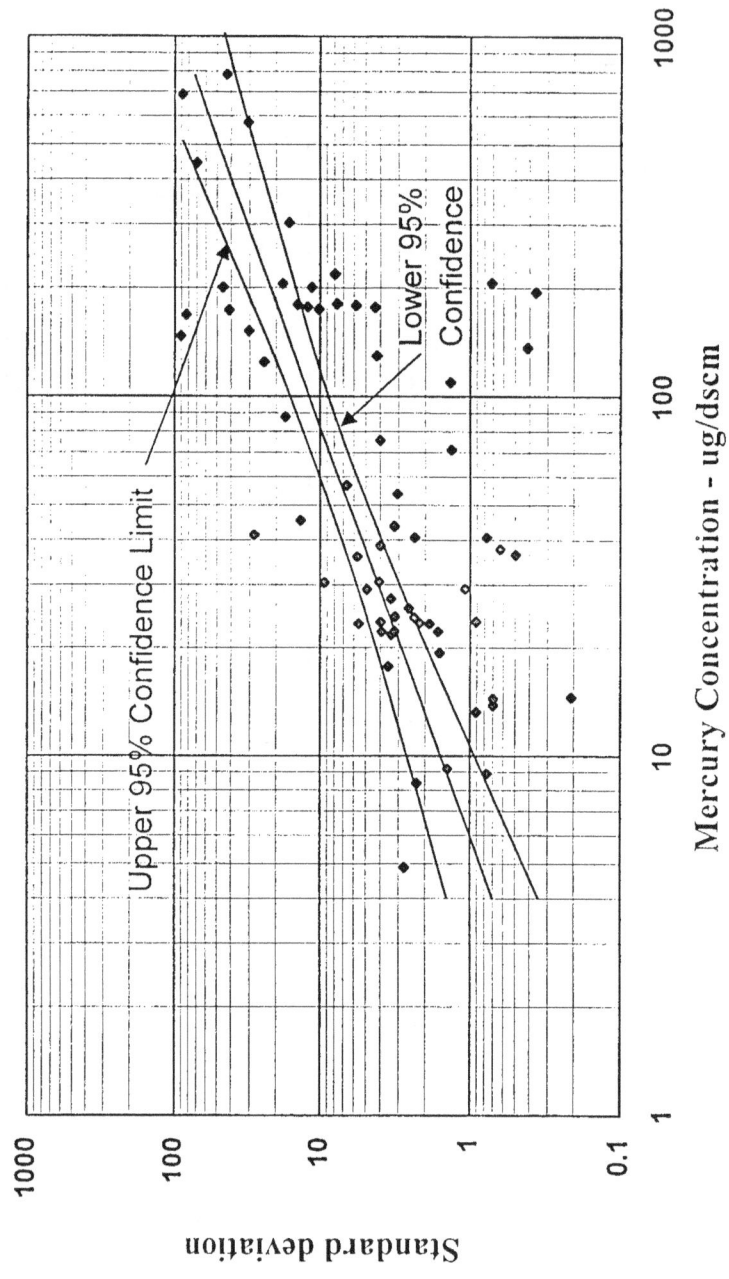

confidence intervals as well as the small sample bias corrected data. The t-statistic for the regression is 7.45, which is well above the critical value for 67 degrees of freedom and 95% confidence. After applying a bias correction factor for the log-log transformation the regression relation is found to be:

$$S \text{ (Hg total, Methods 29 and 101)} = 0.208 * C^{0.877}.$$

The value of the power coefficient, at the 95% confidence level is:

$$P = 0.877 \pm 0.234 \text{ or between } 0.643 \text{ and } 1.112.$$

Figure 38 presents three different forms of the precision for Methods 29 and 101 for measurement of total mercury. The relative standard deviation is a very flat function of concentration, ranging between about 10 and 15%. In fact, the predicted RSD is below 15% at all concentration levels above 10 µg/dscm. Triplicate measurements of total mercury are anticipated to fall within a relatively narrow range of the true stack concentration. For true stack concentrations above 10 µg/dscm, Method imprecision is anticipated to result in 99 out of 100 triplicate measurements deviating by less than ± 23.3% from the actual stack concentration. Similarly, if the true stack concentration is above 10 µg/dscm, 99 out of 100 single measurements can be expected to fall within ± 40.4% of the true stack concentration.

Data presented in Figure 36 are based on the assumption that standard deviation varies with concentration according to the regression relation shown in Figure 37. At the 95% confidence level, it is possible that standard deviation could be as high as the upper confidence interval and as low as the lower confidence interval. The potential range for 99 out of 100 future measurements, due to imprecision in the measurement method, has been calculated under all three scenarios as a function of true stack concentration. Results from those calculations are presented in Figure 39. If the precision of the mercury measurement methods is as large as the upper confidence interval, a significant variation in measurement results may be anticipated. For example, if the method standard deviation varies according to the upper confidence limit, measurement of a stack containing 500 µg/dscm of mercury could result in a spread for 99 out of 100 data points ranging from 275 to 725 µg/dscm. If the Method precision is best described by the lower confidence interval,

114

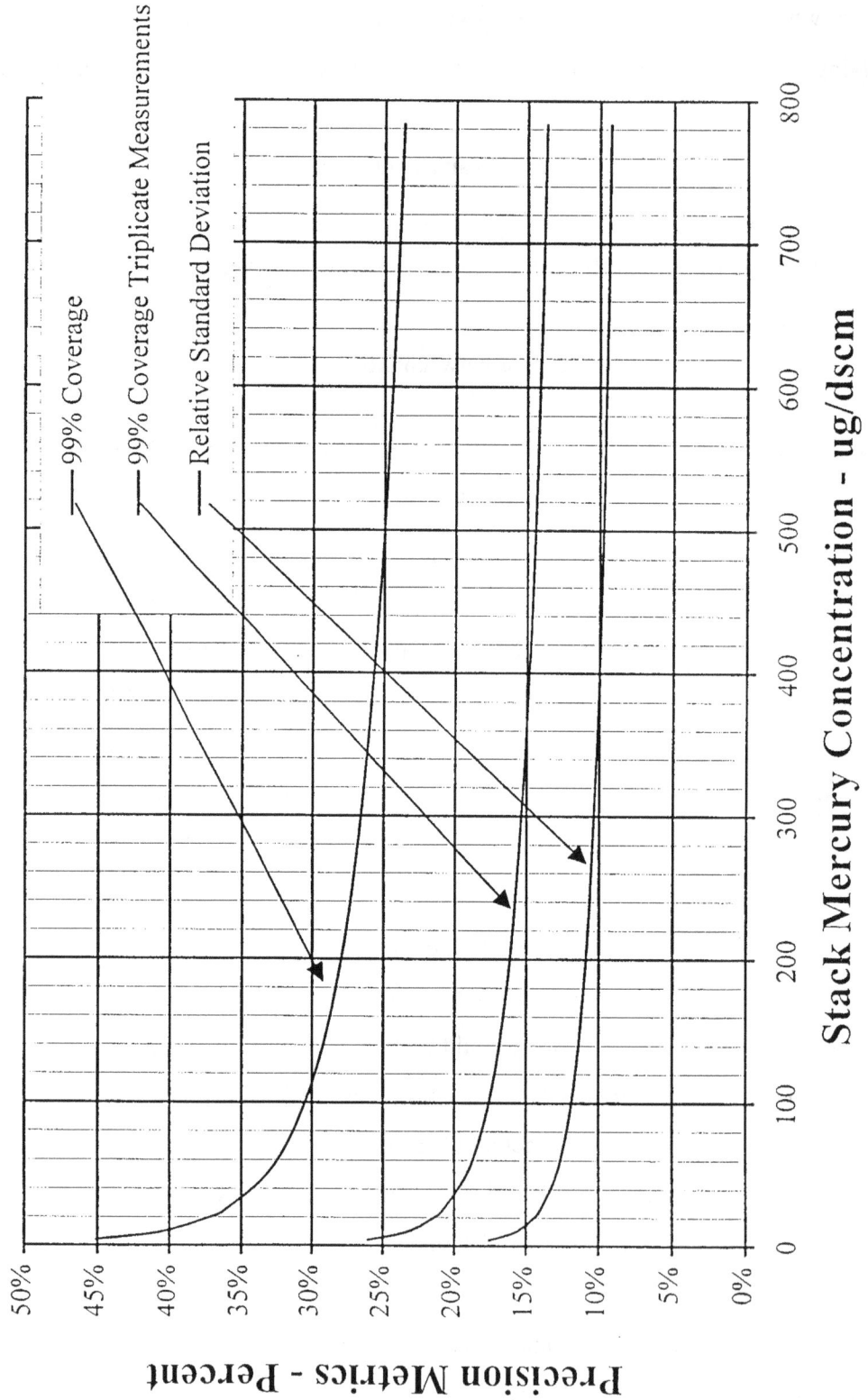

Figure 38. Precision Metrics – EPA Methods 29, 101a and 101b for Total Mercury

Figure 39.   Precision Estimates for Measurements Using EPA
Methods 29 and 101 for Mercury

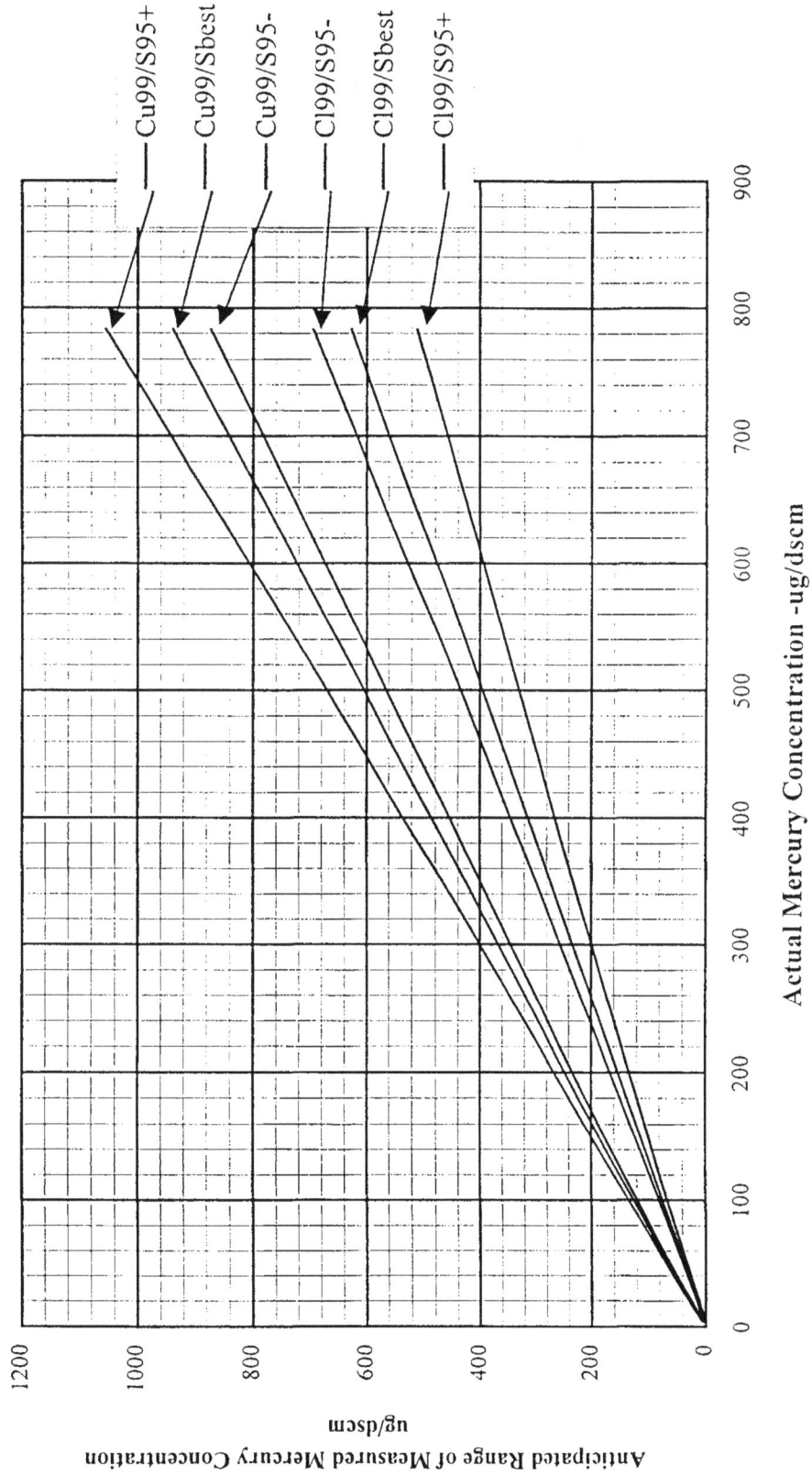

Cu99/S95+
Cu99/Sbest
Cu99/S95-
Cl99/S95-
Cl99/Sbest
Cl99/S95+

Anticipated Range of Measured Mercury Concentration ug/dscm

Actual Mercury Concentration -ug/dscm

measurement of that same stack containing 500 µg/dscm mercury should result in a 99 out of 100 future data points falling between 428 and 572 µg/dscm. It is noteworthy that the RSD data listed in Tables 16a and 16b for individual mercury measurements occasionally exceeded 50% and that the firms conducting the tests were highly qualified stack testers and laboratory analysts. This serves to underline the critical importance of precision in stack testing. From an owner's perspective, it is critical to know the precision of the data being gathered to determine compliance. From a regulatory development perspective, it is critically important to know the precision of data being used to establish regulations.

The best estimate of the precision of Method 29/101A/101B is provided by the regression analysis. Those data were presented graphically in Figures 38 and 39. To assist the reader, Table 17 presents these data in tabular form, focusing on the range of concentrations anticipated to be of primary interest to facilities being regulated under the new and emerging US EPA emission rules. The first column lists the true concentration of total mercury in the stack (not corrected for excess oxygen). The second and third columns present the anticipated range for 99 out of 100 future single measurements (at the given true stack concentration). The fourth and fifth columns provide the anticipated range for 99 out of 100 future triplicate measurements.

Table 17. Range of Anticipated Future Mercury Data

True Stack Total Mercury Concentration ug/dscm	Upper Range For 99 out of 100 Single Measurements	Lower Range For 99 out of 100 Single Measurements	Upper Range For 99 out of 100 Triplicate Measurements
4	5.81	2.19	5.04
10	14.0	5.96	12.3
20	27.4	12.6	24.3
25	34.0	16.0	30.2
50	66.6	33.4	59.6
75	98.7	51.3	88.7
100	130.5	69.5	117.6
125	162.0	88.0	146.4
150	193.5	106.5	175.1
175	224.8	125.2	203.7
200	255.9	144.1	232.3

*   Concentrations are not Oxygen Corrected

** Ranges based on Est. Standard Deviation from Regression Analysis

118

8.0   EPA Method 29 for Multi-Metals

The essential elements of EPA Method 29 were presented in the previous section. As noted, the various components of the train are analyzed to determine the concentration of various metals. The method is used to determine compliance with a wide range of metals but there are very limited multi-train data available to assess the precision of the method for those pollutants. For three of the key metals, cadmium, chromium, and lead, the Agency did perform multi-train, Method validation tests. The various reports from those tests suggest that additional data may have been gathered on other metals but documentation of the results (if any) were not available. Note that lead and cadmium are regulated pollutants under the municipal waste combustor MACT rules. Other metals of current regulatory significance include antimony, arsenic, and beryllium. Other sources of multi-train data include the previously noted tests conducted by Rigo and Chandler as well as the EPA pilot-scale tests, conducted as part of the multi-metals CEM demonstration program.

8.1   EPA Method 29 Data for Antimony, Arsenic, Beryllium, Cadmium, Chromium, and Lead

Tables 18 through 23 summarize the available Method 29 data for antimony, arsenic, beryllium, cadmium, chromium, and lead respectively. In these tables, some difficulty may be experienced relative to the run number designations for data from the EPA pilot-scale tests and for the EPA-ORD/Radian tests (Table 22). Tables 18 through 23 utilize the run number designations provided in the original data source. The difficulty is that the same run numbers are repeated. Separate runs with the same run designation clearly come from different testing series but insufficient information was available to clearly describe the differences between run series or why run designations were repeated.

8.1.1   *Antimony Data* For antimony, data is available from the Rigo and Chandler tests and from the EPA pilot-scale tests. The Rigo and Chandler tests provide dual train data in the 30 to 80 µg/dscm concentration range. The EPA pilot scale tests were also dual train. They provide data in two different concentration ranges; a lower range of approximately 20 to 30 µg/dscm and an upper range of approximately 60 to 90 µg/dscm. The outlier analysis indicates that run number 2 from the

Table 18. Method 29 Multi-Train Data and Standard Deviation For Antimony

Data Source	Run Number	Train A	Train B	Avg Concentration ug/dscm	Standard Deviation	RSD	S - Bias Corrected	RSD - Bias Corrected
EPA Pilot Scale Tests	1	91.68	90.00	90.84	1.19	1.31%	1.49	1.64%
	2	31.39	59.08	45.24	19.58	43.29%	24.54	54.24%
	3	54.51	66.79	60.65	8.68	14.32%	10.88	17.94%
	5	43.76	29.49	36.62	10.09	27.55%	12.64	34.52%
	1	73.72	86.19	79.95	8.81	11.02%	11.04	13.81%
	2	76.10	84.05	80.08	5.62	7.02%	7.05	8.80%
	4	80.22	70.36	75.29	6.97	9.26%	8.74	11.60%
	5	52.24	76.34	64.29	17.05	26.51%	21.36	33.22%
	1	23.74	23.05	23.40	0.49	2.10%	0.62	2.64%
	2	21.35	27.11	24.23	4.07	16.81%	5.10	21.06%
	3	15.75	23.81	19.78	5.70	28.82%	7.14	36.11%
	4	23.80	24.30	24.05	0.35	1.46%	0.44	1.83%
	5	25.03	26.60	25.81	1.11	4.31%	1.39	5.39%
	1	25.94	34.00	29.97	5.70	19.01%	7.14	23.82%
	2	31.07	29.91	30.49	0.82	2.67%	1.02	3.35%
	3	30.03	33.66	31.84	2.57	8.07%	3.22	10.11%
	4	27.04	31.62	29.33	3.24	11.04%	4.06	13.83%
	5	28.26	25.49	26.87	1.96	7.29%	2.45	9.13%
Rigo & Chandler	1	61.99	56.60	59.30	3.81	6.43%	4.78	8.06%
	3	34.82	35.45	35.14	0.45	1.27%	0.56	1.59%
	6	33.51	36.55	35.03	2.15	6.12%	2.69	7.67%
	7	49.82	53.41	51.61	2.54	4.92%	3.18	6.16%
	8	50.43	53.12	51.78	1.90	3.67%	2.38	4.60%
	9	43.43	45.05	44.24	1.14	2.58%	1.43	3.24%
	10	51.83	51.02	51.42	0.58	1.12%	0.72	1.40%
	11	64.40	63.20	63.80	0.85	1.33%	1.06	1.67%
	12	26.09	20.87	23.48	3.69	15.71%	4.62	19.69%
	13	25.04	25.15	25.10	0.08	0.33%	0.10	0.41%
	14	30.10	26.92	28.51	2.25	7.88%	2.82	9.87%
	15	75.87	78.97	77.42	2.19	2.83%	2.74	3.54%
	16	38.52	43.93	41.22	3.82	9.28%	4.79	11.62%
	17	77.29	86.50	81.90	6.51	7.95%	8.16	9.97%
	18	49.87	46.66	48.27	2.26	4.69%	2.84	5.88%
	19	26.25	31.16	28.70	3.47	12.10%	4.35	15.17%

Table 19. Method 29 Multi-Train Data and Standard Deviation For Arsenic

Data Source	Run Number	Train A	Train B	Avg Concentration ug/dscm	Standard Deviation	RSD	S - Bias Corrected	RSD - Bias Corrected
Rigo & Chandler	1	10.24	8.36	9.30	1.33	14.35%	1.67	17.98%
	3	6.16	6.82	6.49	0.46	7.16%	0.58	8.97%
	6	4.64	5.48	5.06	0.60	11.76%	0.75	14.74%
	7	10.75	12.02	11.38	0.90	7.87%	1.12	9.86%
	8	7.03	7.44	7.24	0.29	3.94%	0.36	4.94%
	9	3.83	4.00	3.92	0.12	3.11%	0.15	3.89%
	10	7.77	7.40	7.59	0.27	3.52%	0.33	4.41%
	11	6.71	6.32	6.51	0.27	4.22%	0.34	5.28%
	12	3.13	2.19	2.66	0.67	25.00%	0.83	31.32%
	13	3.62	3.83	3.72	0.15	4.01%	0.19	5.03%
	14	6.29	5.65	5.97	0.45	7.58%	0.57	9.49%
	15	7.99	8.95	8.47	0.68	8.04%	0.85	10.08%
	16	6.33	6.86	6.60	0.38	5.74%	0.47	7.19%
	17	7.20	8.13	7.66	0.66	8.58%	0.82	10.76%
	18	5.25	4.67	4.96	0.41	8.31%	0.52	10.41%
	19	2.23	2.60	2.41	0.26	10.71%	0.32	13.42%
EPA Pilot Scale Tests	1	88.06	87.22	87.64	0.59	0.67%	0.74	0.84%
	2	43.11	62.36	52.74	13.61	25.81%	17.05	32.34%
	3	59.83	83.18	71.50	16.51	23.09%	20.69	28.93%
	5	48.48	32.66	40.57	11.19	27.57%	14.02	34.55%
	1	76.69	86.75	81.72	7.11	8.70%	8.91	10.90%
	2	78.48	90.96	84.72	8.83	10.42%	11.06	13.06%
	4	86.25	77.28	81.76	6.34	7.76%	7.95	9.72%
	5	57.97	78.66	68.31	14.63	21.41%	18.33	26.83%
	1	26.71	26.80	26.76	0.06	0.24%	0.08	0.30%
	2	23.25	21.98	22.62	0.90	3.97%	1.12	4.97%
	3	15.79	25.61	20.70	6.94	33.54%	8.70	42.03%
	4	26.20	25.17	25.69	0.73	2.84%	0.91	3.55%
	5	25.56	28.70	27.13	2.22	8.17%	2.78	10.24%
	1	23.84	33.07	28.46	6.53	22.94%	8.18	28.74%
	2	29.31	28.20	28.75	0.78	2.73%	0.98	3.42%
	3	27.96	31.12	29.54	2.24	7.57%	2.80	9.49%
	4	26.10	30.87	28.49	3.37	11.85%	4.23	14.84%
	5	28.42	17.63	23.02	7.63	33.12%	9.55	41.50%

Table 20. Method 29 Multi-Train Data and Standard Deviation For Beryllium

Data Source	Run Number	Train A	Train B	Avg Concentration ug/dscm	Standard Deviation	RSD	S - Bias Corrected	RSD - Bias Corrected
	1	83.23	80.00	81.62	2.29	2.80%	2.86	3.51%
	2	32.77	52.46	42.62	13.92	32.67%	17.45	40.93%
	3	50.62	71.83	61.23	15.00	24.50%	18.80	30.70%
	5	39.27	25.81	32.54	9.52	29.26%	11.93	36.66%
	1	69.56	78.89	74.22	6.60	8.89%	8.26	11.13%
	2	69.56	80.60	75.08	7.81	10.40%	9.78	13.03%
	4	71.77	64.59	68.18	5.08	7.45%	6.36	9.33%
	5	51.82	67.09	59.46	10.80	18.16%	13.53	22.76%
Pilot Pilot Scale Tests	1	18.96	19.71	19.33	0.54	2.77%	0.67	3.47%
	2	16.65	15.35	16.00	0.92	5.78%	1.16	7.24%
	3	11.26	18.46	14.86	5.10	34.29%	6.38	42.97%
	4	19.31	18.39	18.85	0.65	3.43%	0.81	4.29%
	5	18.24	20.72	19.48	1.75	9.01%	2.20	11.29%
	1	18.93	25.51	22.22	4.65	20.94%	5.83	26.23%
	2	23.99	22.16	23.07	1.30	5.63%	1.63	7.06%
	3	22.26	24.75	23.51	1.77	7.51%	2.21	9.41%
	4	19.85	22.82	21.33	2.10	9.83%	2.63	12.32%
	5	21.31	13.38	17.35	5.61	32.32%	7.02	40.49%

Table 21. Method 29 Multi-Train Data and Standard Deviation For Cadmium

Data Source	Run Number	Train A	Train B	Train C	Train D	Average Concentration ug/dscm	Standard Deviation	RSD	S - Bias Corrected	RSD - Bias Corrected
Rigo & Chandler	1	43.13	45.82			44.47	1.90	4.28%	2.39	5.36%
	3	25.72	30.00			27.86	3.03	10.87%	3.80	13.62%
	6	28.36	36.55			32.45	5.79	17.84%	7.26	22.36%
	7	44.57	45.40			44.99	0.58	1.29%	0.73	1.62%
	8	47.78	50.47			49.12	1.90	3.87%	2.38	4.85%
	9	22.99	24.02			23.51	0.73	3.11%	0.91	3.89%
	10	33.69	30.61			32.15	2.18	6.77%	2.73	8.49%
	11	42.94	42.14			42.54	0.57	1.33%	0.71	1.67%
	12	19.83	17.52			18.68	1.63	8.73%	2.04	10.94%
	13	24.20	27.07			25.64	2.03	7.90%	2.54	9.90%
	14	22.99	21.54			22.26	1.02	4.60%	1.28	5.76%
	15	31.95	36.85			34.40	3.47	10.09%	4.35	12.64%
	16	22.84	24.16			23.50	0.94	3.98%	1.17	4.99%
	17	39.98	41.94			40.96	1.39	3.39%	1.74	4.25%
	18	31.49	28.52			30.01	2.11	7.02%	2.64	8.79%
	19	17.59	20.77			19.18	2.25	11.75%	2.82	14.73%
EPA Pilot Scale Tests	1	83.23	80.56			81.89	1.89	2.31%	2.37	2.90%
	2	41.19	57.99			49.59	11.88	23.95%	14.88	30.01%
	3	57.11	80.03			68.57	16.21	23.63%	20.31	29.61%
	5	45.52	31.33			38.42	10.03	26.11%	12.57	32.72%
	1	74.32	83.38			78.85	6.41	8.13%	8.03	10.18%
	2	74.91	86.36			80.63	8.09	10.04%	10.14	12.58%
	4	83.84	74.39			79.12	6.68	8.44%	8.37	10.57%
	5	56.90	78.08			67.49	14.98	22.20%	18.77	27.81%
	1	22.44	23.16			22.80	0.51	2.24%	0.64	2.81%
	2	19.14	17.65			18.39	1.05	5.73%	1.32	7.18%
	3	13.29	19.38			16.33	4.31	26.39%	5.40	33.06%
	4	20.65	19.65			20.15	0.70	3.50%	0.88	4.38%
	5	20.27	23.12			21.69	2.02	9.30%	2.53	11.65%
	1	20.18	27.90			24.04	5.46	22.70%	6.84	28.44%
	2	24.70	22.45			23.58	1.59	6.73%	1.99	8.43%
	3	23.74	26.27			25.01	1.79	7.15%	2.24	8.96%
	4	21.19	25.33			23.26	2.93	12.59%	3.67	15.77%
	5	22.99	14.96			18.98	5.68	29.91%	7.11	37.48%
EPA/OAQPS Radian	1		3.01	3.27	0.99	2.42	1.24	51.40%	1.40	57.97%
	2	1.34	3.58	3.70	2.29	2.73	1.12	41.19%	1.22	44.70%
	3	0.91	2.77	3.31		2.33	1.26	54.15%	1.42	61.08%
	4	1.38	3.64	3.48	2.13	2.66	1.09	40.93%	1.18	44.41%
	5	1.98	2.79	2.87	0.55	2.05	1.07	52.38%	1.16	56.83%
	6		1.84	1.89	0.88	1.53	0.57	37.11%	0.64	41.86%
	7	1.24	2.02	2.12	1.23	1.65	0.49	29.36%	0.53	31.85%
	8	0.68	2.56	1.62	0.66	1.38	0.91	65.68%	0.98	71.26%

Table 22. Method 29 Multi-Train Data and Standard Deviation For Chromium

Data Source	Run Number	Train A	Train B	Train C	Train D	Average Concentration ug/dscm	Standard Deviation	RSD	S - Bias Corrected	RSD - Bias Corrected
Rigo & Chandler	1	8.09	9.97			9.03	1.33	14.8%	1.67	18.5%
	3	6.70	12.00			9.35	3.75	40.1%	4.70	50.3%
	6	5.41	7.05			6.23	1.16	18.6%	1.45	23.2%
	7 ****	16.78	40.06			28.42	16.46	57.9%	20.62	72.6%
	8	8.10	7.70			7.90	0.28	3.5%	0.35	4.4%
	9	5.11	5.76			5.43	0.46	8.4%	0.57	10.5%
	10	7.77	7.40			7.59	0.27	3.5%	0.33	4.4%
	11	8.05	16.85			12.45	6.23	50.0%	7.80	62.6%
	12	4.44	3.35			3.89	0.77	19.7%	0.96	24.7%
	13	2.64	5.19			3.92	1.80	46.0%	2.26	57.7%
	14	4.93	7.54			6.23	1.85	29.6%	2.31	37.1%
	15	4.26	3.95			4.10	0.22	5.4%	0.28	6.7%
	16	4.95	4.67			4.81	0.20	4.2%	0.25	5.3%
	17	11.99	6.29			9.14	4.03	44.1%	5.05	55.3%
	18	6.04	6.22			6.13	0.13	2.1%	0.16	2.7%
	19	4.46	3.12			3.79	0.95	25.1%	1.19	31.5%
EPA Pilot Scale Tests	1	70.57	72.78			71.67	1.56	2.2%	1.96	2.7%
	2	33.97	55.80			44.89	15.43	34.4%	19.34	43.1%
	3	52.16	76.24			64.20	17.03	26.5%	21.34	33.2%
	5	45.58	34.34			39.96	7.95	19.9%	9.96	24.9%
	1	61.83	75.52			68.68	9.68	14.1%	12.13	17.7%
	2	65.99	77.72			71.86	8.29	11.5%	10.39	14.5%
	4	74.79	66.90			70.84	5.58	7.9%	6.99	9.9%
	5	50.51	75.77			63.14	17.86	28.3%	22.38	35.4%
	1	28.02	29.68			28.85	1.18	4.1%	1.47	5.1%
	2	25.39	24.28			24.84	0.78	3.1%	0.98	3.9%
	3	18.70	25.65			22.17	4.92	22.2%	6.16	27.8%
	4	27.54	27.65			27.59	0.07	0.3%	0.09	0.3%
	5	29.41	33.20			31.30	2.68	8.5%	3.35	10.7%
	1	22.13	31.14			26.63	6.37	23.9%	7.98	30.0%
	2	29.46	26.89			28.18	1.82	6.5%	2.28	8.1%
	3	27.49	32.37			29.93	3.45	11.5%	4.33	14.5%
	4	26.38	32.22			29.30	4.13	14.1%	5.18	17.7%
	5	29.08	19.66			24.37	6.66	27.3%	8.34	34.2%
EPA ORD-Entropy	3	2.60	4.00	3.20		3.27	0.70	21.5%	0.79	24.3%
	5		3.60	3.90		3.75	0.21	5.7%	0.27	7.1%
	8	1.60	1.80	2.20		1.87	0.31	16.4%	0.34	18.5%
	10	2.30	2.10	1.90		2.10	0.20	9.5%	0.23	10.7%
	3	5.60	8.60	4.30	6.10	6.15	1.80	29.3%	1.95	31.8%
	7	4.10	3.90	4.40		4.13	0.25	6.1%	0.28	6.9%
	9	3.10	3.60			3.35	0.35	10.6%	0.44	13.2%
	11	3.80	3.40			3.60	0.28	7.9%	0.35	9.8%
	13	2.70	2.30			2.50	0.28	11.3%	0.35	14.2%
	6	1.40	0.80			1.10	0.42	38.6%	0.53	48.3%
	3	11.80	10.70	12.40		11.63	0.86	7.4%	0.97	8.4%
	5	17.20	15.20	15.30		15.90	1.13	7.1%	1.27	8.0%
	8 ****	17.50	18.30	4.20		13.33	7.92	59.4%	8.93	67.0%
	10	14.50	15.30	14.60		14.80	0.44	2.9%	0.49	3.3%
	4	3.00	1.40	1.70		2.03	0.85	41.8%	0.96	47.2%
	6	1.50	2.90	1.50		1.97	0.81	41.1%	0.91	46.4%
	8	1.20	1.50	1.30		1.33	0.15	11.5%	0.17	12.9%

Table 23. Method 29 Multi-Train Data and Standard Deviation For Lead

Data Source	Run Number	Train A	Train B	Train C	Train D	Avg Concentration ug/dscm	Standard Deviation	RSD	S - Bias Corrected	RSD - Bias Corrected
Rigo & Chandler	1	1617.25	1428.49			1522.87	133.47	8.76%	167.24	10.98%
	3	910.77	981.81			946.29	50.23	5.31%	62.94	6.65%
	6	644.46	704.81			674.64	42.67	6.33%	53.47	7.93%
	7	1022.58	1014.74			1018.66	5.54	0.54%	6.94	0.68%
	8	955.57	1009.32			982.44	38.01	3.87%	47.63	4.85%
	9	510.93	550.56			530.75	28.02	5.28%	35.11	6.62%
	10	1088.45	1020.33			1054.39	48.17	4.57%	60.35	5.72%
	11	1288.07	1211.39			1249.73	54.22	4.34%	67.94	5.44%
	12	626.21	541.18			583.69	60.12	10.30%	75.33	12.91%
	13	862.43	902.26			882.34	28.16	3.19%	35.29	4.00%
	14	903.00	834.59			868.79	48.37	5.57%	60.61	6.98%
	15	1011.63	1079.26			1045.45	47.82	4.57%	59.92	5.73%
	16	770.39	823.66			797.03	37.66	4.73%	47.19	5.92%
	17	1332.59	1389.29			1360.94	40.09	2.95%	50.24	3.69%
	18	866.10	829.60			847.85	25.81	3.04%	32.34	3.81%
	19	393.71	441.44			417.58	33.75	8.08%	42.29	10.13%
EPA Pilot Scale Tests	1	93.49	93.33			93.41	0.11	0.12%	0.14	0.14%
	2	42.09	62.36			52.23	14.33	27.44%	17.96	34.39%
	3	61.80	87.59			74.70	18.23	24.41%	22.84	30.58%
	5	49.94	34.06			42.00	11.23	26.74%	14.07	33.50%
	1	89.18	98.26			93.72	6.42	6.85%	8.04	8.58%
	2	89.77	100.75			95.26	7.76	8.15%	9.72	10.21%
	4	98.31	85.35			91.83	9.16	9.98%	11.48	12.50%
	5	68.06	89.65			78.85	15.26	19.36%	19.13	24.26%
	1	27.54	28.13			27.83	0.41	1.49%	0.52	1.86%
	2	23.49	23.04			23.26	0.32	1.36%	0.40	1.70%
	3	15.12	25.27			20.19	7.18	35.55%	8.99	44.55%
	4	26.60	25.32			25.96	0.90	3.46%	1.13	4.34%
	5	26.15	29.51			27.83	2.38	8.55%	2.98	10.71%
	1	26.65	35.08			30.86	5.97	19.33%	7.48	24.22%
	2	30.95	27.83			29.39	2.21	7.52%	2.77	9.42%
	3	30.65	32.37			31.51	1.22	3.86%	1.52	4.84%
	4	27.04	32.03			29.54	3.53	11.95%	4.42	14.97%
	5	29.47	17.56			23.51	8.42	35.83%	10.55	44.89%
EPA ORD Radian	1	45.56	44.25	44.60	44.65	44.76	0.56	1.25%	0.61	1.35%
	2	28.47	41.48	41.08	34.35	36.34	6.19	17.02%	6.71	18.47%
	3	30.05	34.78	41.23	31.38	34.36	5.00	14.54%	5.42	15.78%
	4	17.84	46.31	46.79	47.66	39.65	14.55	36.70%	15.79	39.82%
	5	44.65	38.07	41.97	47.69	43.10	4.08	9.47%	4.43	10.28%
	6	7.39	23.17	23.92	19.20	18.42	7.64	41.46%	8.29	44.98%
	7	21.43	19.77	23.25	26.20	22.66	2.75	12.14%	2.98	13.17%
	8	23.16	25.57	15.94	26.84	22.88	4.87	21.28%	5.28	23.08%

EPA pilot scale tests (first of EPA pilot-tests designated as run 2) has an abnormally large data range. After examining several data grouping, it was concluded that this run represents a data outlier and accordingly, it was eliminated from the precision analysis. Figures 40 and 41 present the antimony data in graphical form. Figure 40 is a scatter plot of the small-sample bias corrected standard deviation data while Figure 41 is a scatter plot of the bias corrected relative standard deviation. As shown in Figure 41, the RSD values range from very low levels to approximately 30%.

8.1.2   *Arsenic Data*   Multi-train data sources for arsenic are also limited to results from the Rigo and Chandler tests and from the EPA pilot scale tests. The Rigo and Chandler data are in the concentration range of about 2 to 11 μg/dscm. The EPA pilot scale data fall into two groups centered at about 20 and at about 80 μg/dscm. Outlier analysis suggests that all the arsenic data points should be included in the precision analysis. Figures 42 and 43 present scatter plots of the small sample bias corrected standard deviation and relative standard deviation data. As shown in Figure 43, the data fall into three concentration groups. The relative standard deviation for each group ranges from very low levels to RSD's as high as about 30%.

8.1.3   *Beryllium Data*   Multi-train data for beryllium is the most limited of all Method 29 metals. Data were collected during the Rigo and Chandler tests on the Pittsfield MWC but all of the results indicated concentrations below the analytical detection limit for the laboratory. The only available multi-train data are results from the EPA pilot-scale tests. Those data are in two concentration groups with a lower group range of about 10 – 20 μg/dscm and a higher concentration group in the range of about 60 - 80 μg/dscm. All these data points pass the SPC outlier criteria and have been included in the precision analysis. Figures 44 and 45 present scatter plots of the small sample bias corrected standard deviation and relative standard deviation as a function of the average beryllium test point concentration. Figure 45 shows that the RSD for each group ranges from very low levels to high values of approximately 45%.

8.1.4   *Cadmium Data* Multi-train data for cadmium emissions using method 29 are available from three sources. Tests by Rigo and Chandler provide data in the concentration range of about 20 to 50 μg/dscm. The EPA pilot scale tests provide data in two ranges; a low range centered at about 20

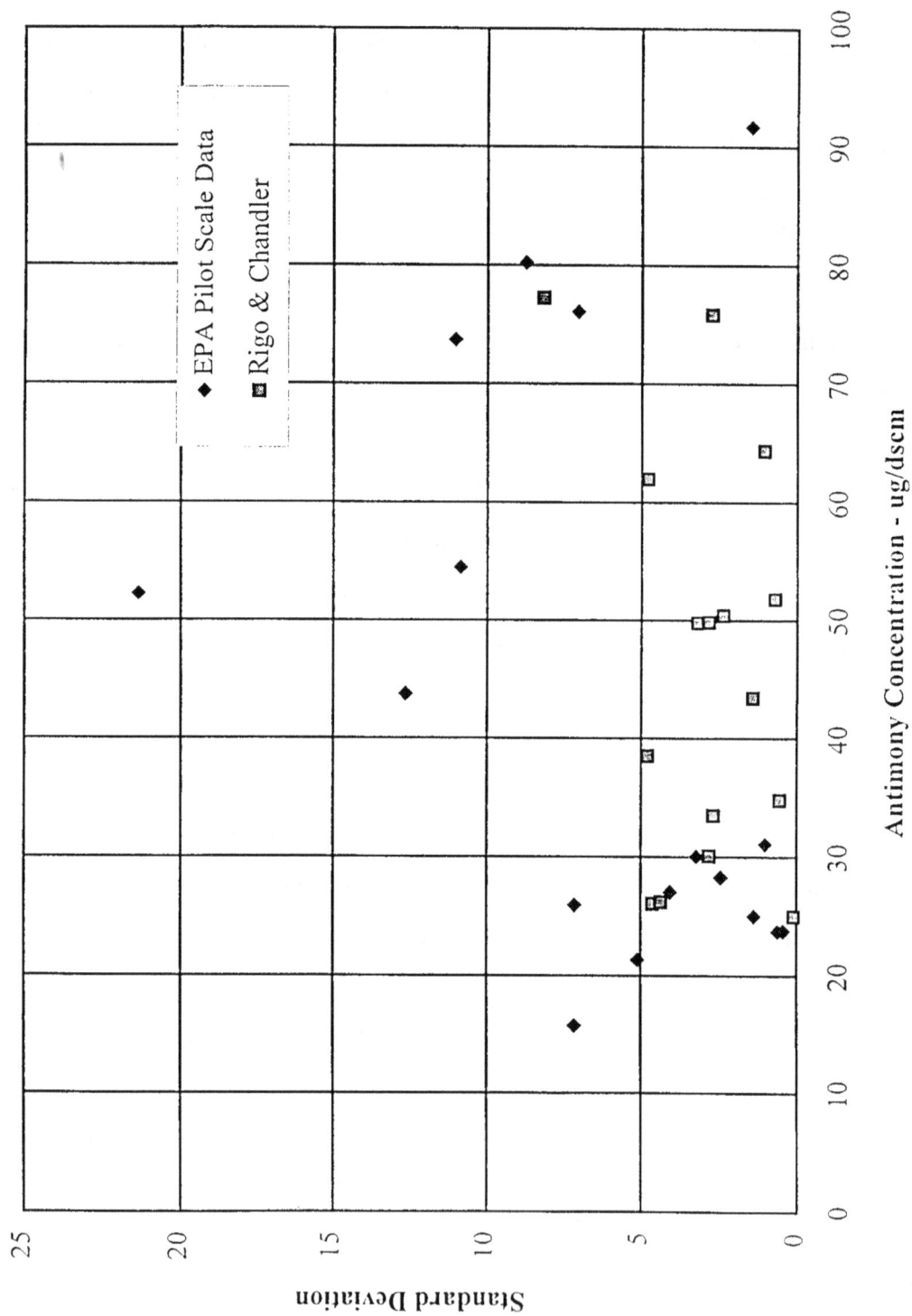

Figure 40. EPA Method 29 Data For Antimony - Standard Deviation

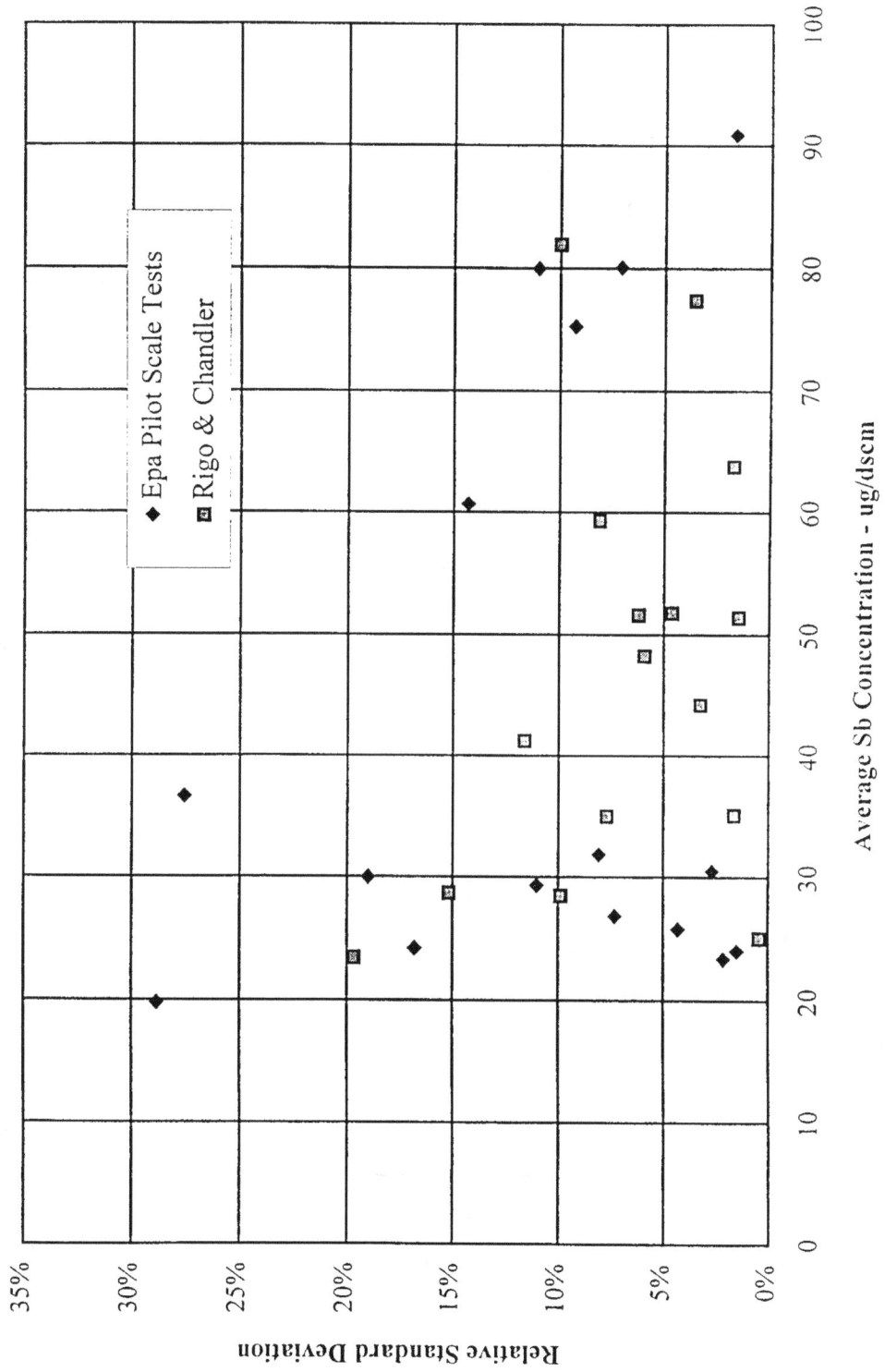

Figure 41. EPA Method 29 Data For Antimony
Relative Standard Deviation

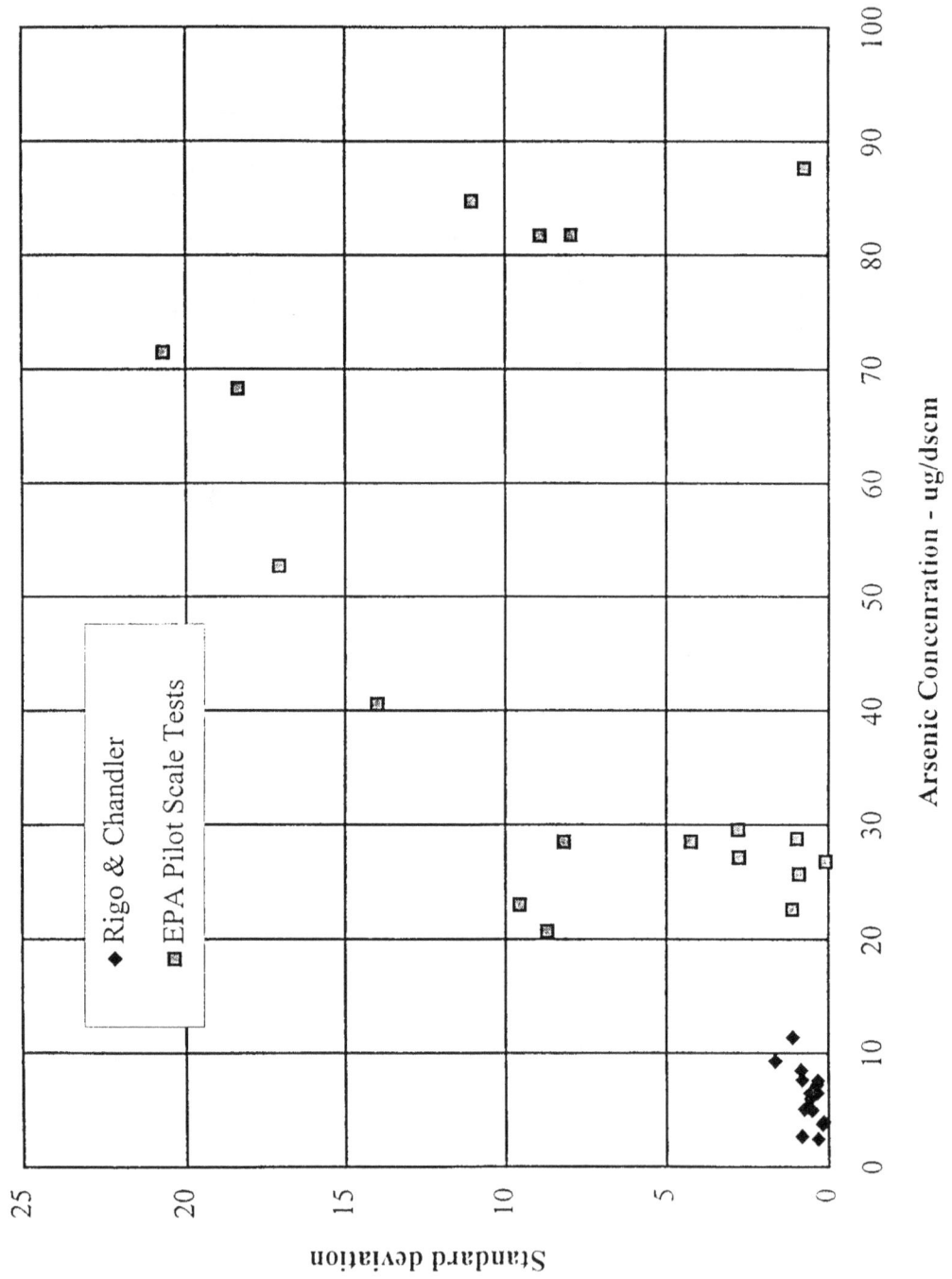

Figure 42. EPA Method 29 Data for Arsenic- Standard Deviation

Figure 43. EPA Method 29 Data for Arsenic Relative Standard Deviation

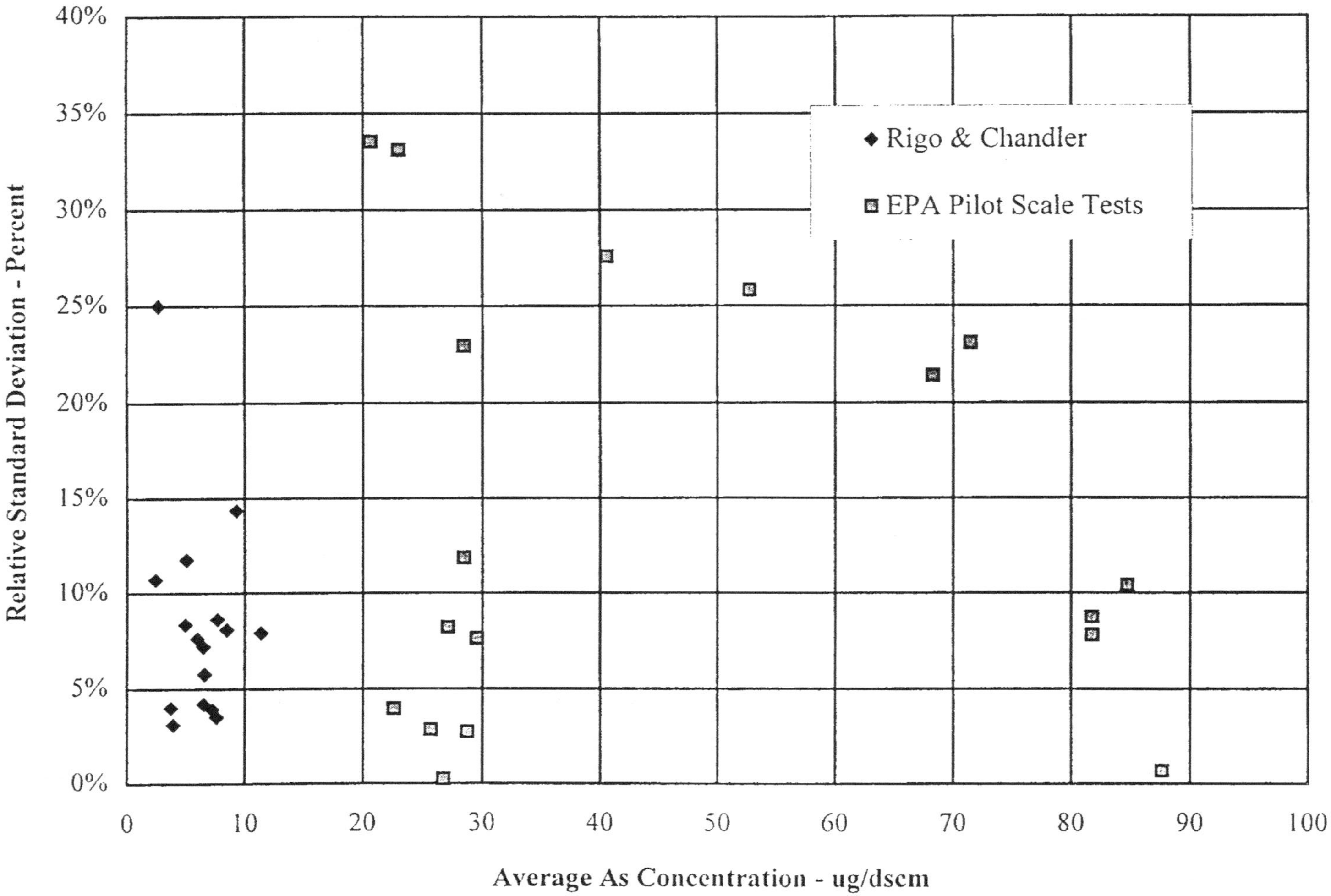

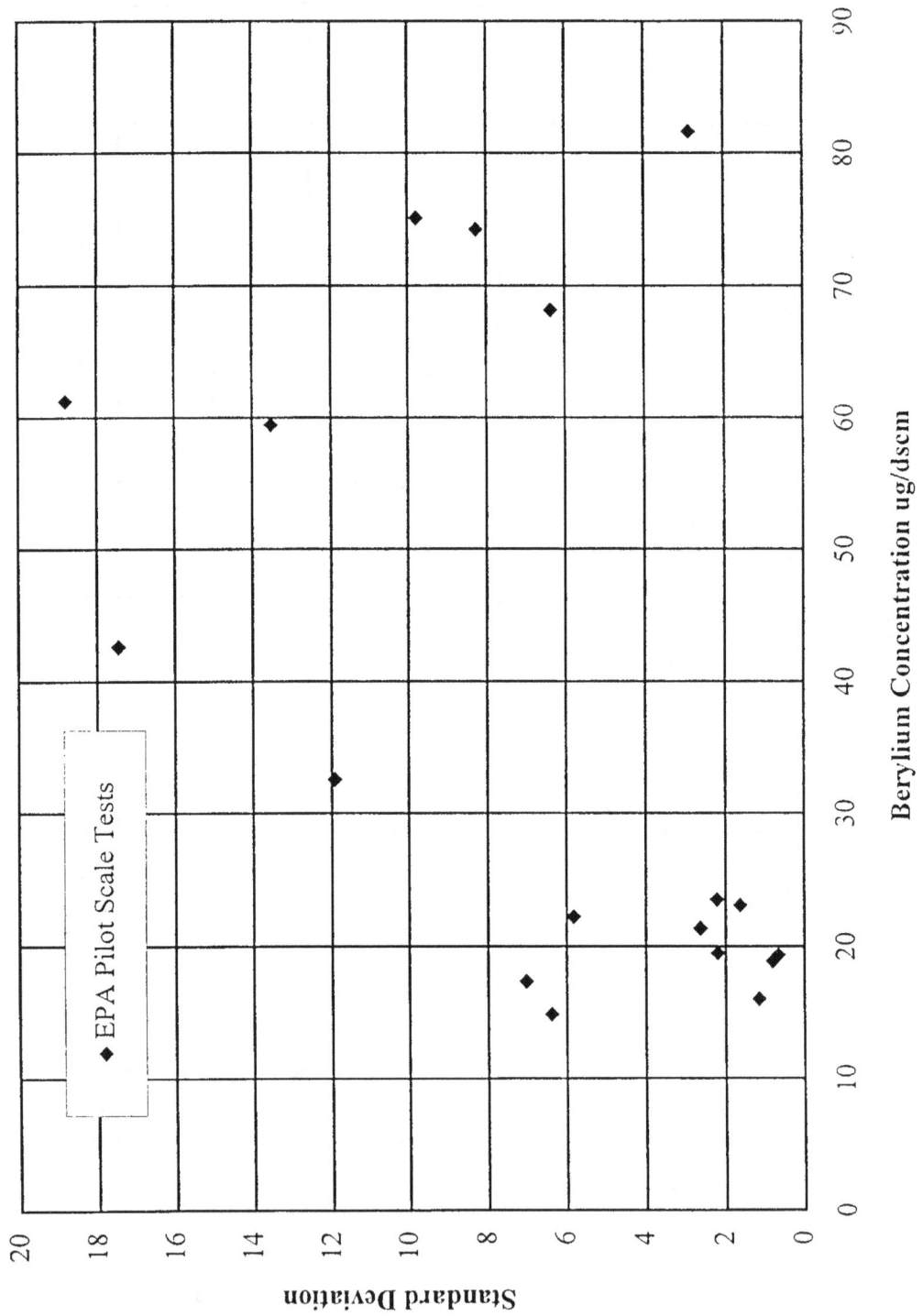

Figure 44. EPA Method 29 Data For Beryllium - Standard Deviation

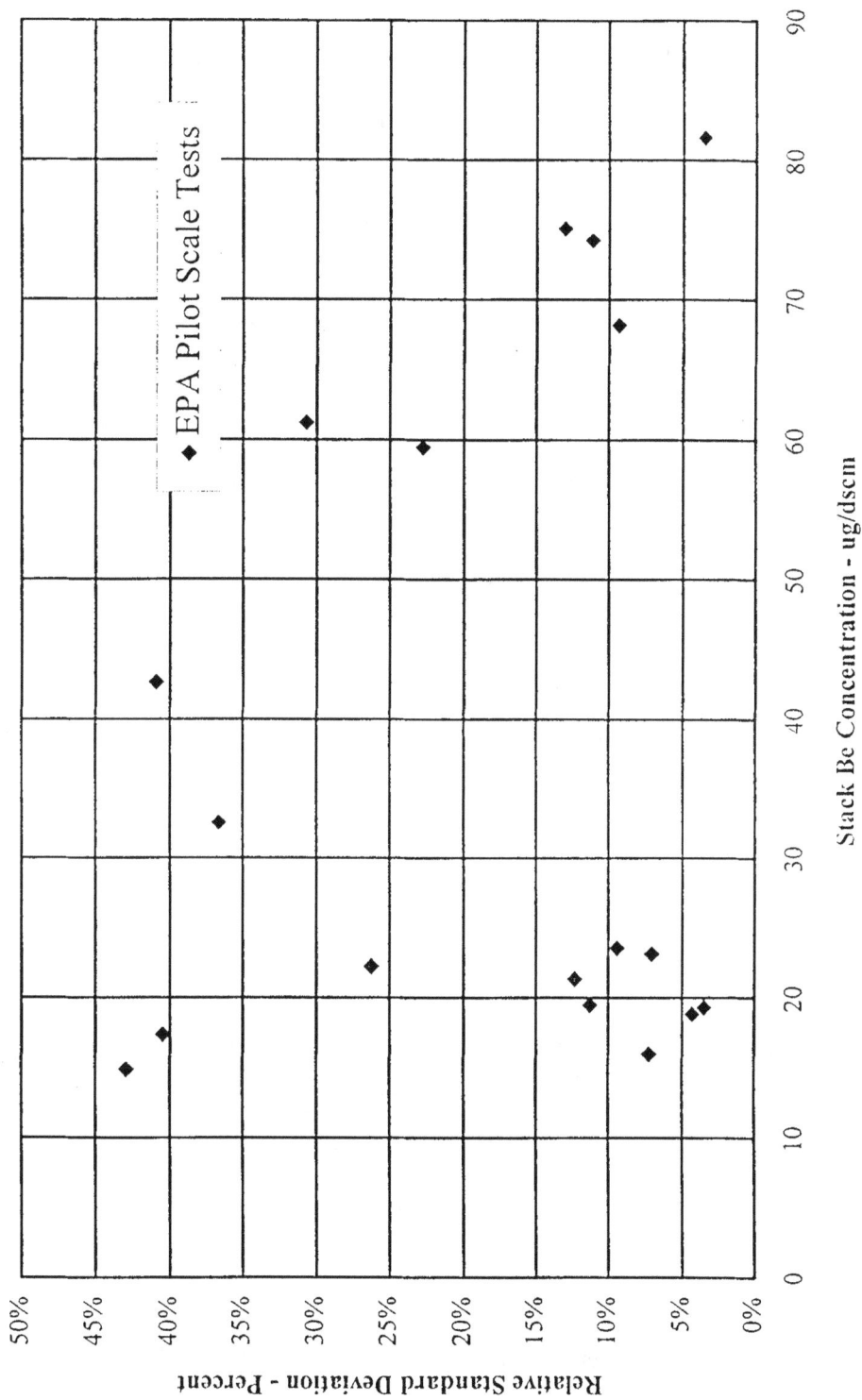

Figure 45. EPA Method 29 Data For Beryllium
Relative Standard Deviation

µg/dscm and an upper range that spans 40 to 80 µg/dscm. The EPA method validation tests provide data at a very low concentration range of about 1.5 to 2.5 µg/dscm. All data in this data set pass the SPC outlier criteria and have been included in the precision analysis. Figures 46 and 47 present scatter plots of the data including the small sample bias corrected standard deviation and relative standard deviation as a function of run average concentration. Figure 47 shows that, for concentrations above about 20 µg/dscm, the RSD covers approximately the same span as the other Method 29 metals presented above. However, the low concentration data from the Method validation tests indicate significantly higher RSD. Data in Figure 46 show that the actual standard deviation tends to increase with increasing concentration, even at the very low concentrations of the validation tests. This shows that the significant rise in RSD is a result of the denominator in the RSD calculation tending toward zero rather than rapid expansion of the standard deviation.

8.1.5 *Chromium Data*   Three sets of multi-train data are available for assessment of the precision of Method 29 for chromium. The Rigo and Chandler tests provide data with concentration in the range of about 4 to 10 µg/dscm. The EPA pilot scale tests again provide data in two ranges. The low range results are centered at about 20 µg/dscm while the high concentration results cover the range of about 60 to 70 µg/dscm. Finally, the EPA Method validation tests are at very low cadmium concentration – ranging from about 1 to 3 µg/dscm. Outlier analysis for this data set indicates that two data points have abnormally large data spreads. Specifically, run number 7 from the Rigo and Chandler tests and run number 8 for the EPA method validation tests. Both of these runs are specially marked in the run number column of Table 22. The Rigo and Chandler run was eliminated from further analysis, while only the data from train C of the EPA tests were eliminated. Note that there were several other data points indicating very large spreads but they have all been included in the analysis. Figures 48 and 49 present scatter plots of the small sample bias corrected data for standard deviation and relative standard deviation as a function of run average concentration.

8.1.6 *Lead Data*   Three sets of multi-train data are also available for Method 29 measurement of lead. The Rigo and Chandler tests at the Pittsfield MWC provide data at lead concentrations ranging from about 400 to 1500 µg/dscm. The EPA pilot scale tests provide data in two concentration ranges.   The low range data cover the span of about 20 to 30 µg/dscm

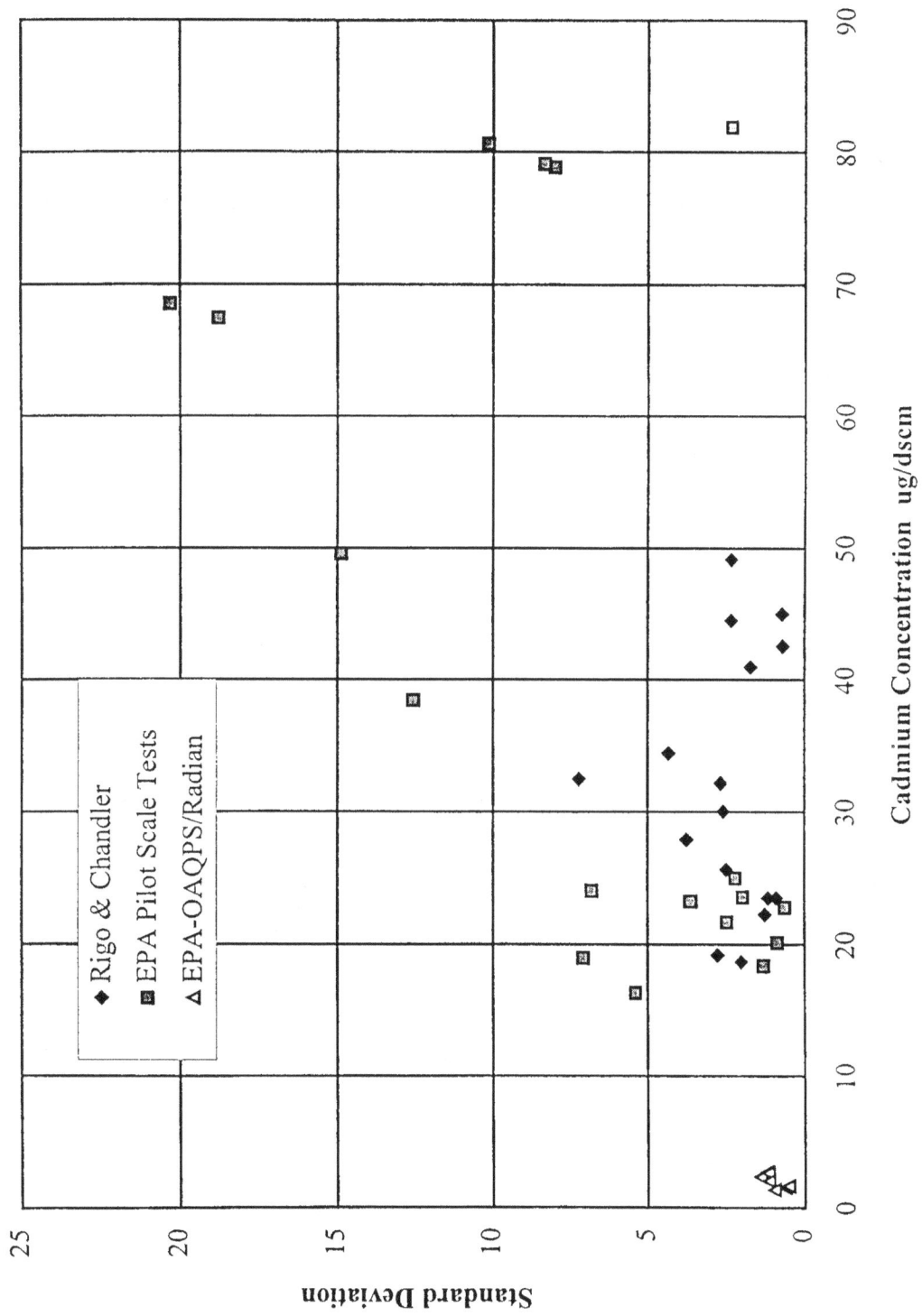

Figure 46. EPA Method 29 Data for Cadmium - Standard Deviation

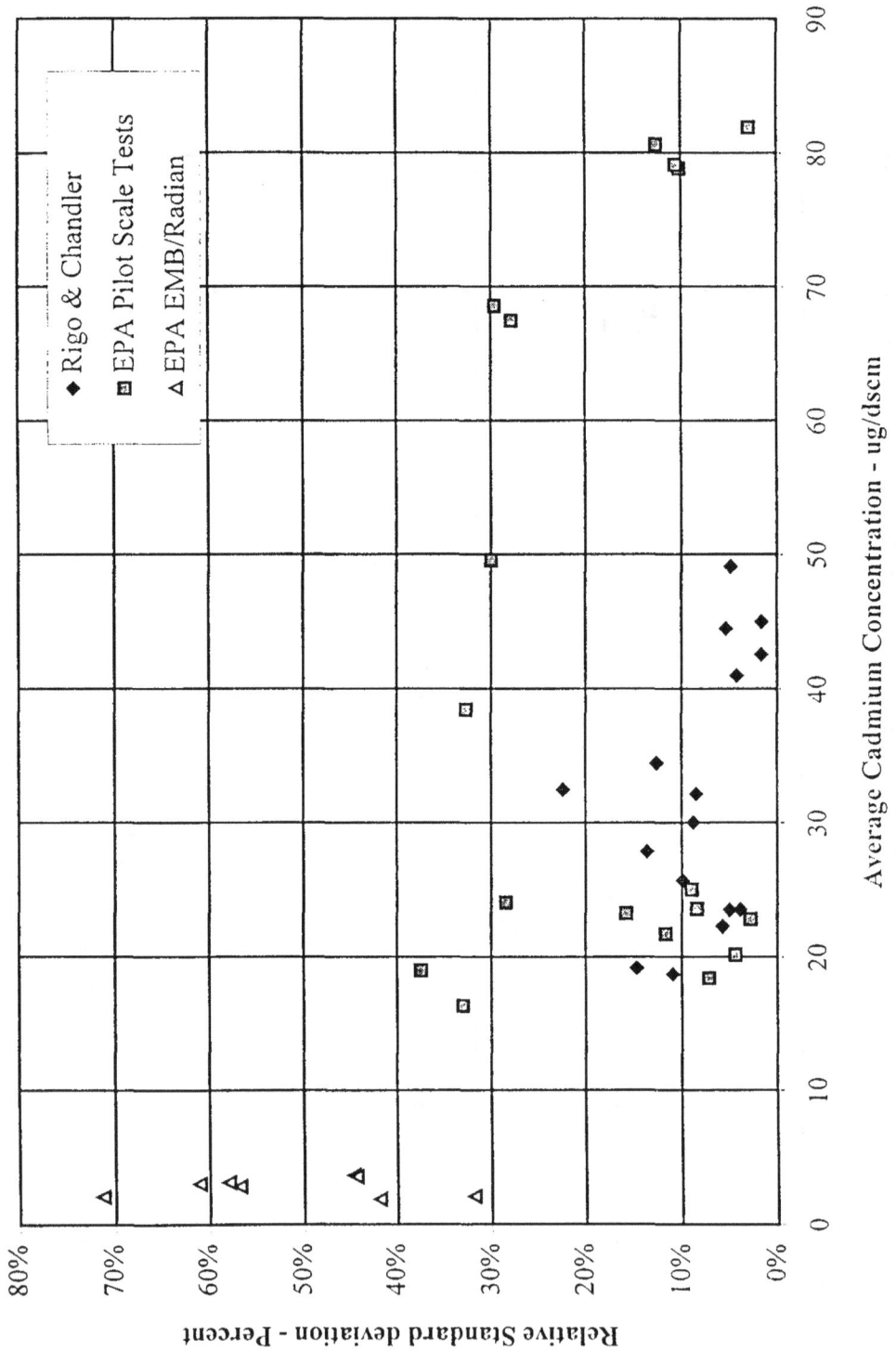

Figure 47. EPA Method 29 Data for Cadmium
Relative Standard Deviation

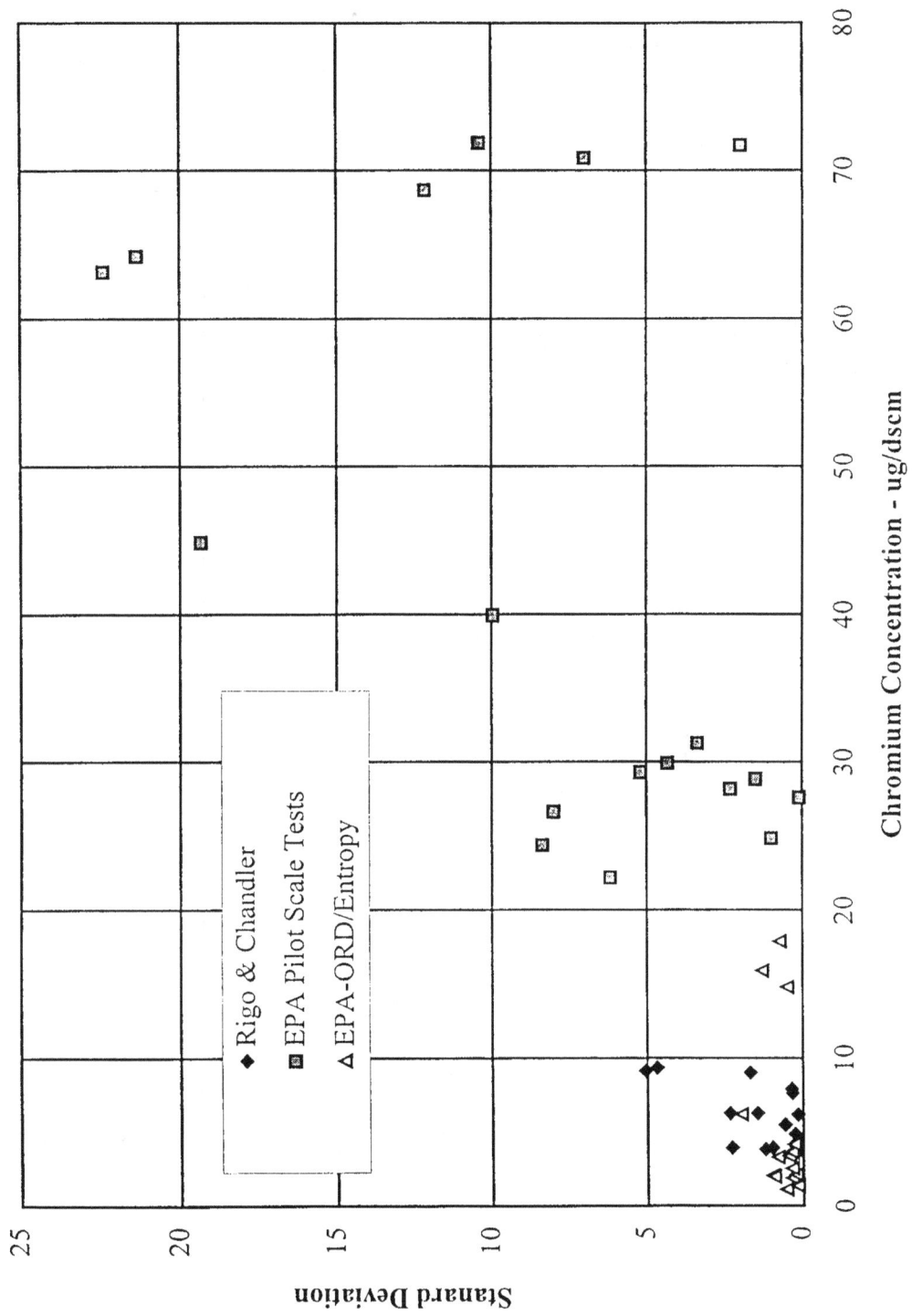

Figure 48. EPA Method 29 Data for Chromium - Standard Deviation

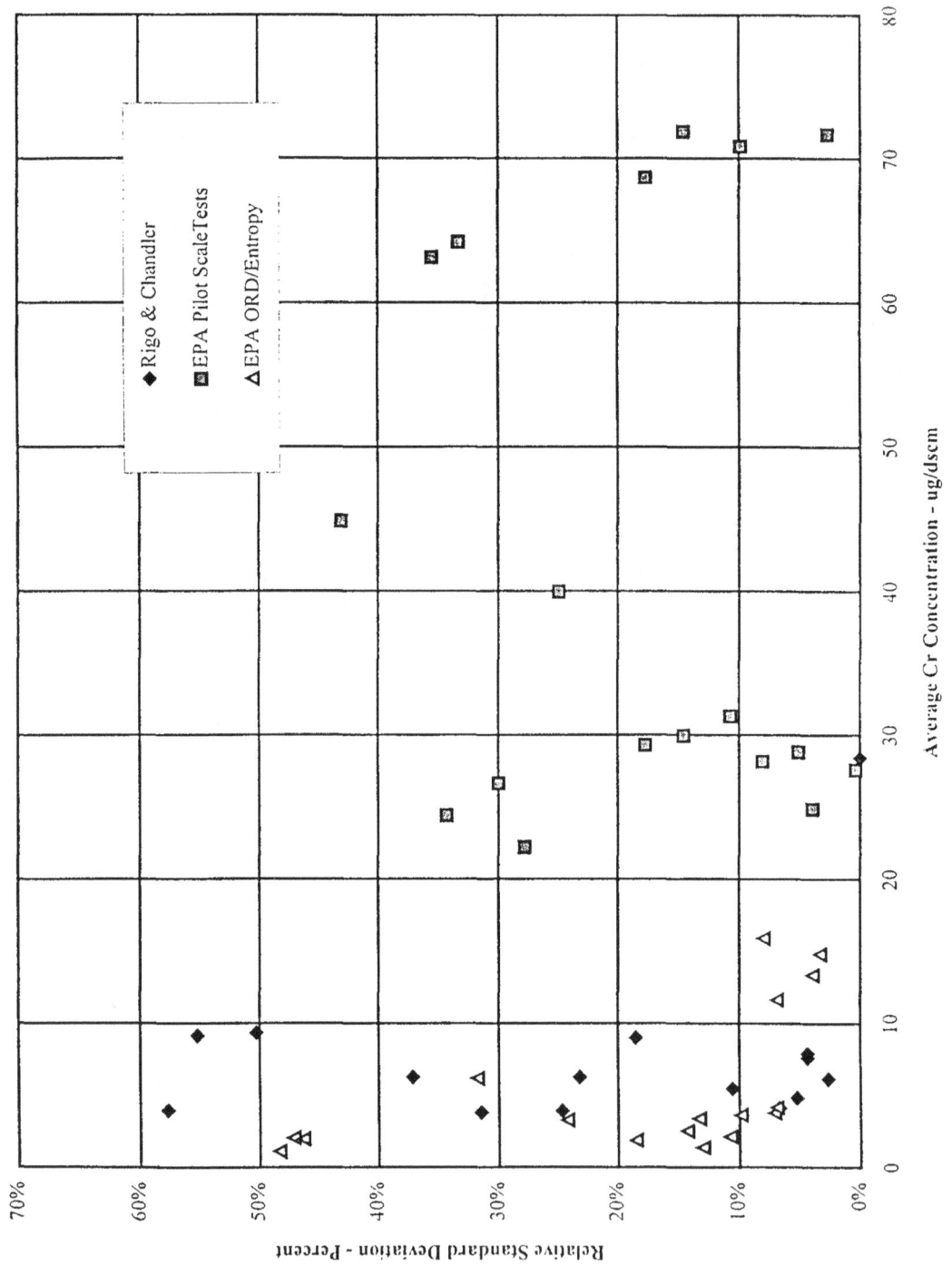

Figure 49. EPA Method 29 Data for Chromium
Relative Standard Deviation

137

while the higher concentration results were obtained in the range of about 50 to 95 μg/dscm. Finally the EPA Method validation test results were obtained at concentrations in the range of 20 to 40 μg/dscm. All data in the lead data set pass the SPC outlier criteria. Figures 50 and 51 present scatter plots of the small sample bias corrected standard deviation and relative standard deviation versus run average concentration. Note that the high concentration data from the Rigo and Chandler tests all exhibit RSD values less than 13%, while the lower concentration results from the two sets of EPA data exhibit a broader range of RSD values.

8.2   EPA Method 29 Regression Analyses

After elimination of outliers, the Method 29 multi-train data for antimony, arsenic, beryllium, cadmium, chromium, and lead were analyzed with weighted regression analysis. Results from those analyses are summarized in Table 24 below.

Table 24.   Results of Method 29 Regression Analysis for Various Metals

Metal	N - Degrees of Freedom	t-Statistic	Log-Log Bias Correction	K	p	95% Confidence Range for p
Antimony	31	2.01	1.514	0.188	0.843	-0.001 to 1.687
Arsenic	32	6.13	1.602	0.136	1.039	0.692 to 1.383
Beryllium	16	2.83	1.292	0.191	0.973	0.244 to 1.702
Cadmium	40	5.73	1.443	0.978	0.452	0.293 to 0.611
Chromium	47	6.97	1.662	0.344	0.833	0.593 to 1.073
Lead	40	6.29	1.293	0.480	0.703	0.577 to 0.929

There are clearly differences between the regression equations for the six metals listed above but there are also significant similarities. The variation of standard deviation for cadmium versus concentration appears to be significantly different from the other metals. For cadmium, the maximum potential value for the p term in the power function, at the 95% confidence level, is only 0.611. For any of the other five metals, within the 95% confidence bounds, the value of the power coefficient could fall between 0.692 and 0.929. Five of the six regression analyses produced t-

Figure 50. EPA Method 29 Data For Lead - Standard Deviation

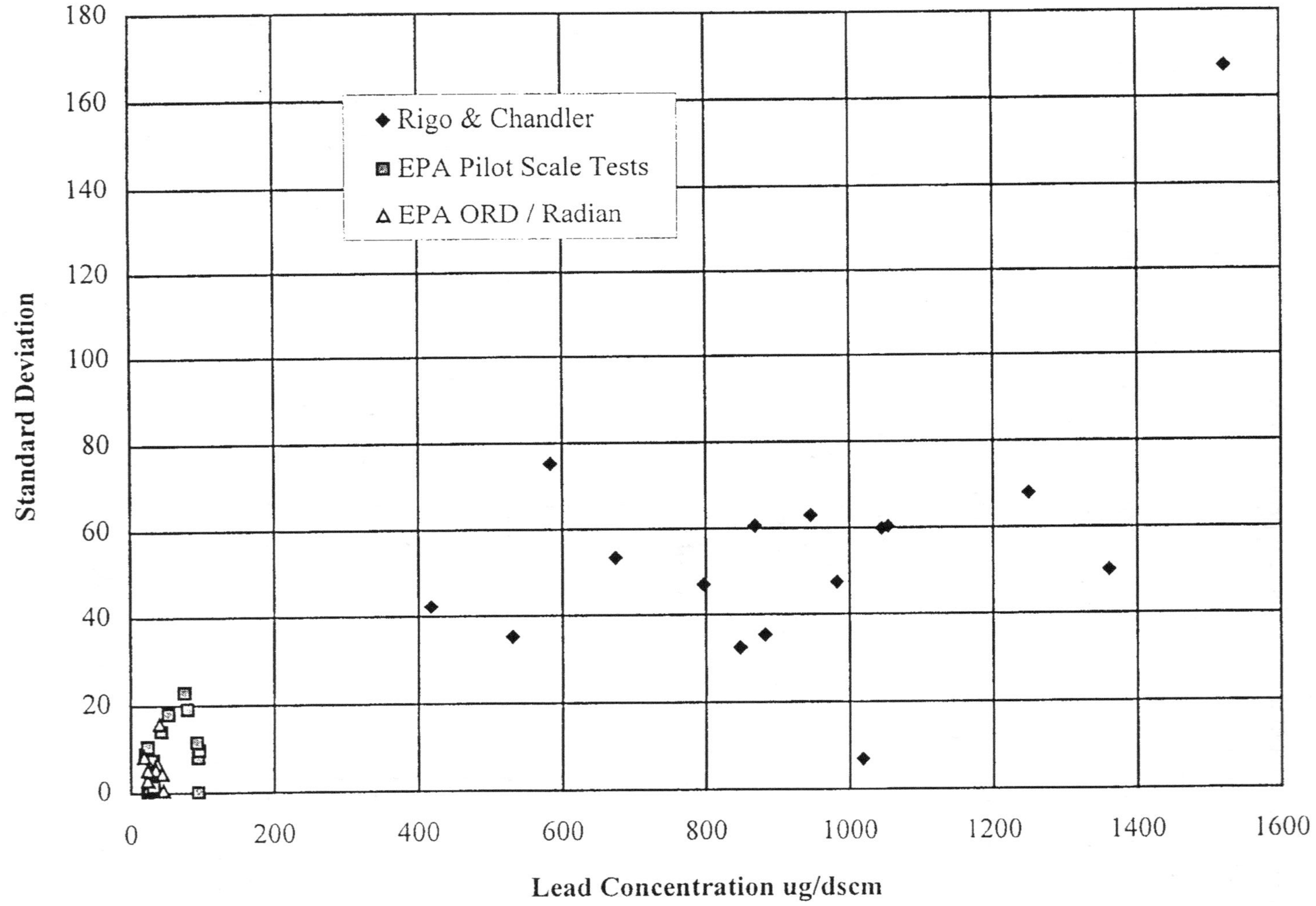

Figure 51. EPA Method 29 Data For Lead
Relative Standard Deviation

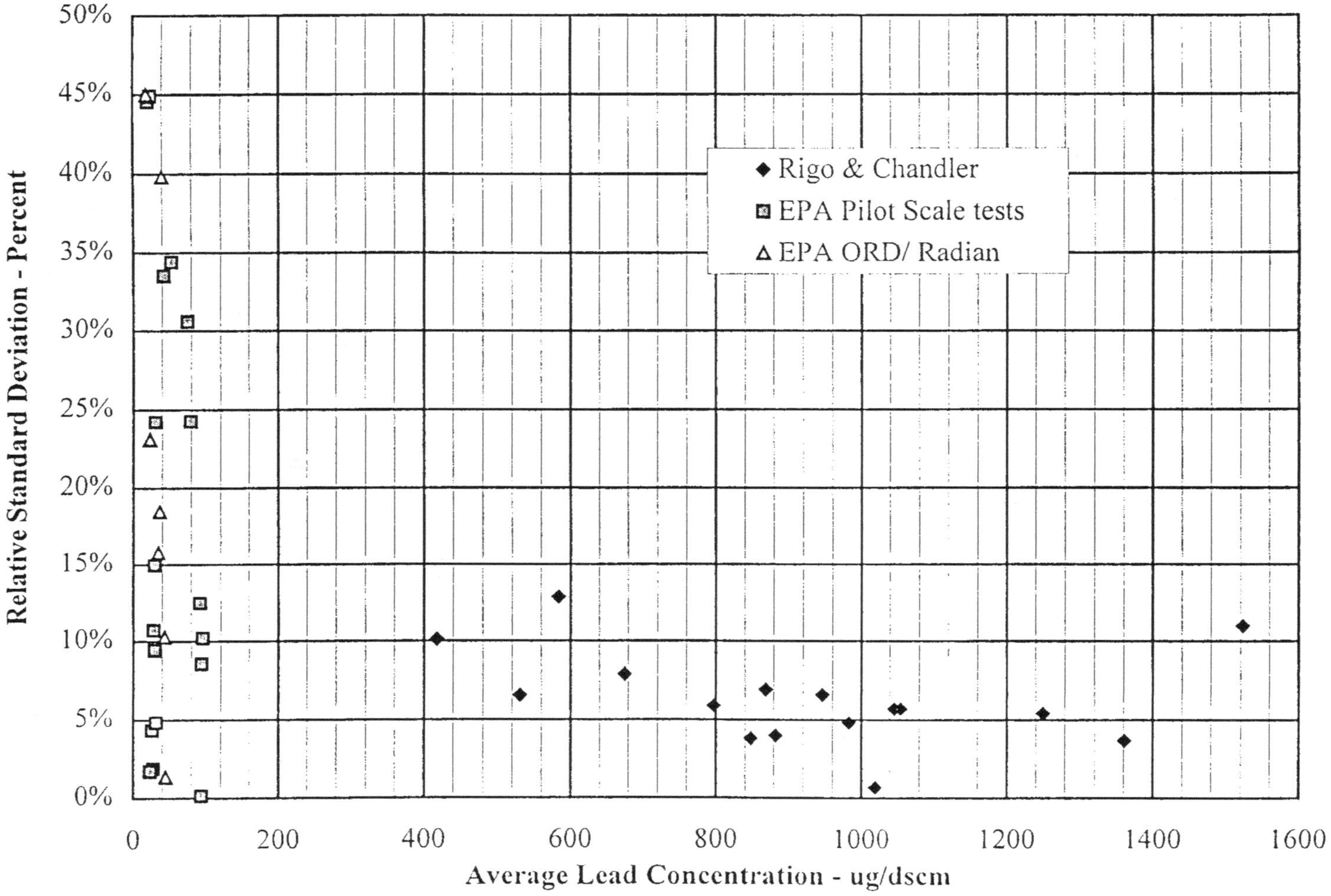

statistics that are above the critical value of that statistic. For antimony the t-statistics (at the 95% confidence level) is slightly below the critical value and the t-statistic for beryllium is only marginally above the critical values. This results in wide ranges for the potential values of the power function p coefficient for both antimony and beryllium.

Recall that, with Method 29, each data pair provides data on the full range of metals. With the exception of the EPA Method validation tests, every data point, for each metal, has a companion data point for the other five metals. This is important since random error enters the measurement process through both the sampling process and the chemical analysis processes. In a multi-metal procedure such as Method 29, random errors in the sampling process should be reflected in every metal being monitored. Moreover, when the value of the power coefficient is close to 1.0, random error in the sample collection process is a significant contributor to the overall Method precision. For these reasons, there is increased reason to anticipate similarity in the various relationships between standard deviation and concentration.

Figures 52 through 57 illustrate the regression equations and the 95% confidence intervals for measurement of each of the six metals using Method 29. Figures 58 through 63 present data on the various precision metrics, assuming that the standard deviation varies according to the regression equation. It is instructive to compare the general level for the predicted relative standard deviation and the variation of RSD for each of the six metals. Figure 58 indicates that the RSD for antimony measurement using Method 29 is basically a flat function of concentration over the entire range of available data. At a true concentration of 20 µg/dscm the predicted RSD is 11.8% while the predicted RSD is 9.3% at 90µg/dscm. In the concentration range between about 20 to 100µg/dscm, a relatively flat RSD versus concentration trend is also observed for arsenic, beryllium, and chromium. In that range, the RSD for arsenic varies between about 15 and 16.3% (see Figure 59). For beryllium, RSD only varies between 17.0% and 17.8% (see Figure 60). Slightly greater variation is predicted for chromium but, as shown in Figure 62, the anticipated variation in RSD is only from 21.3% at 20 µg/dscm to 18.2% at 71 µg/dscm.

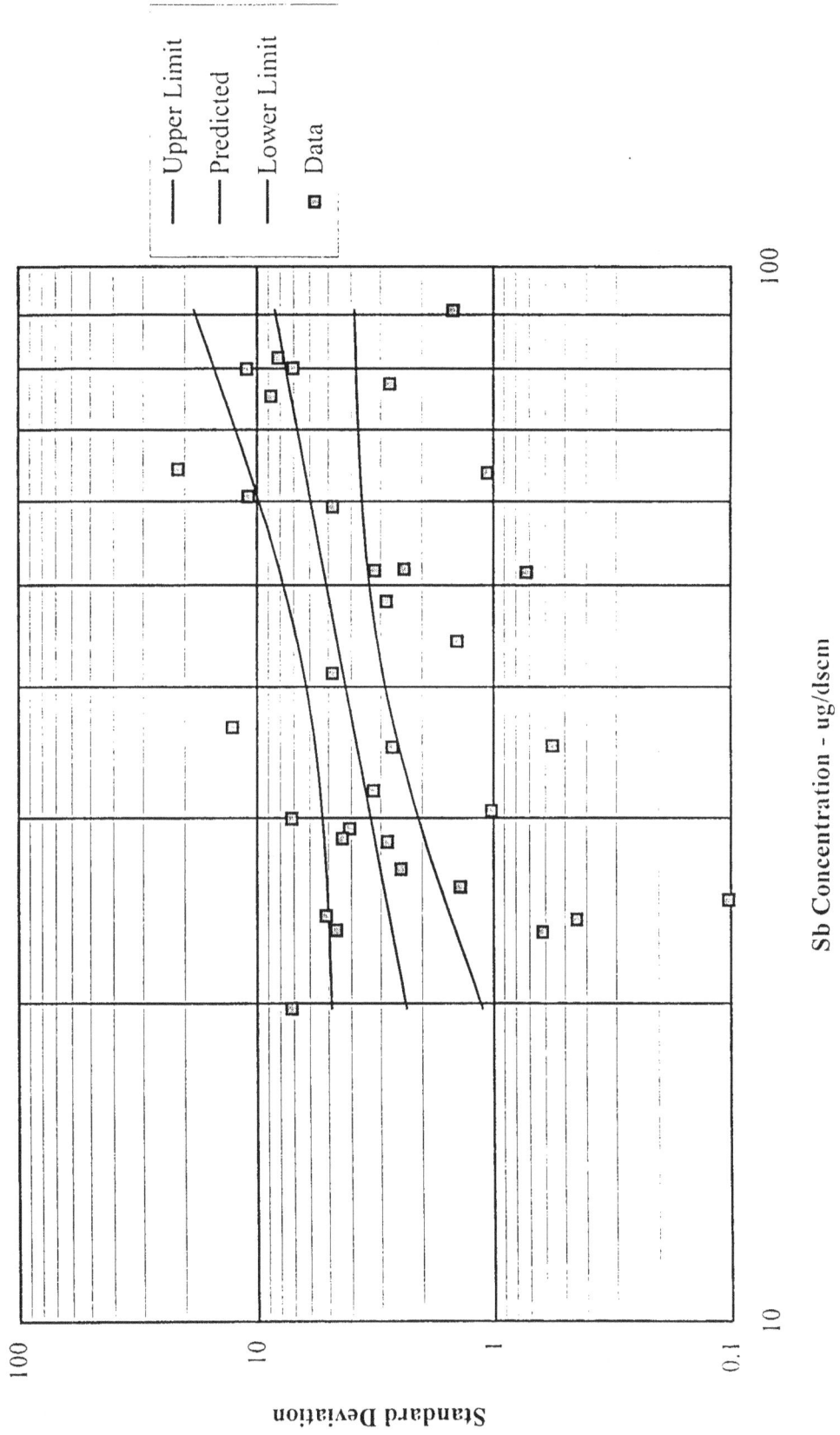

Figure 52. Regression Line and 95% Confidence Interval for EPA Method 29 - Antimony

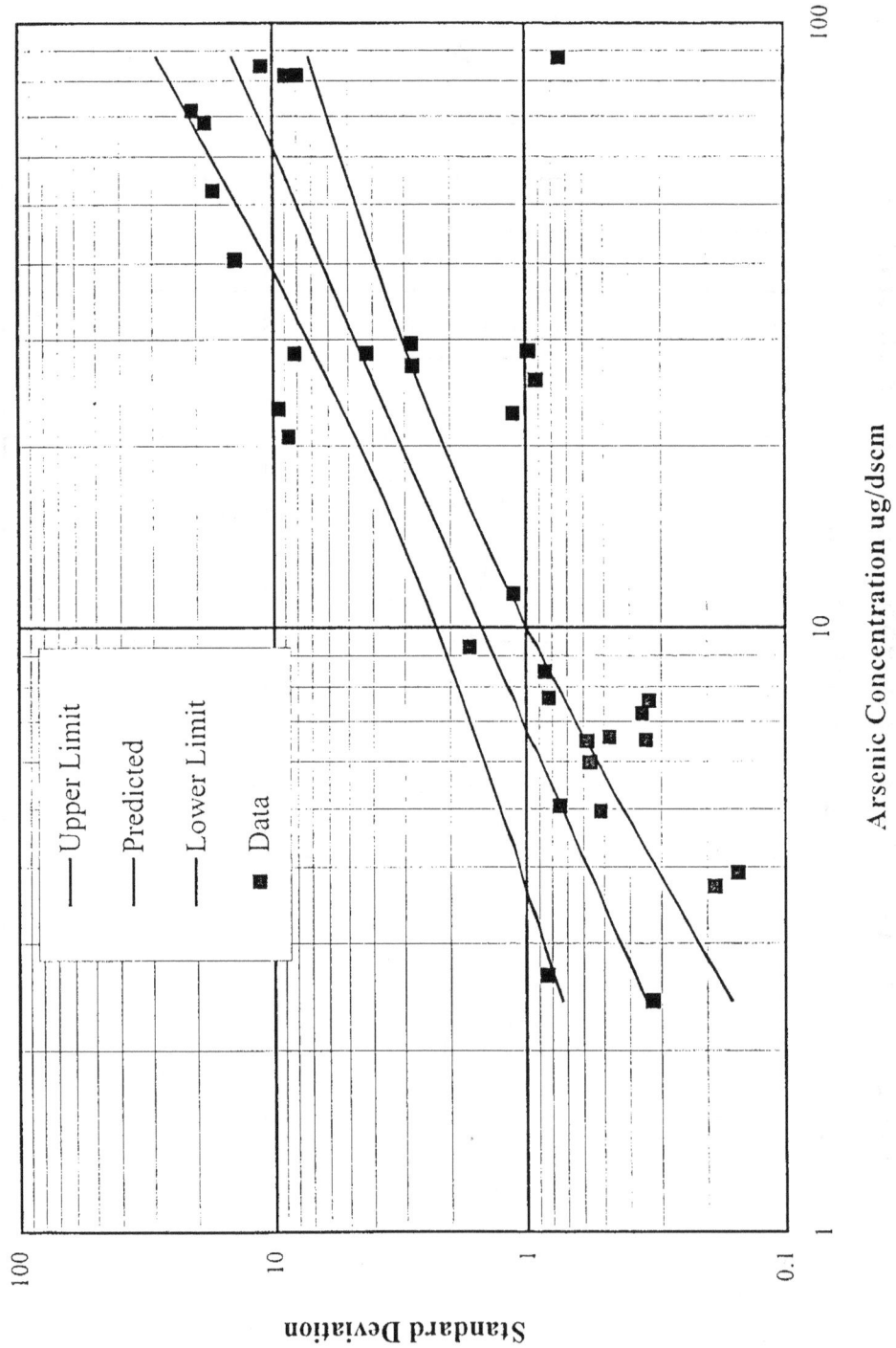

Figure 53. Regression Line and 95% Confidence Interval for EPA Method 29 - Arsenic

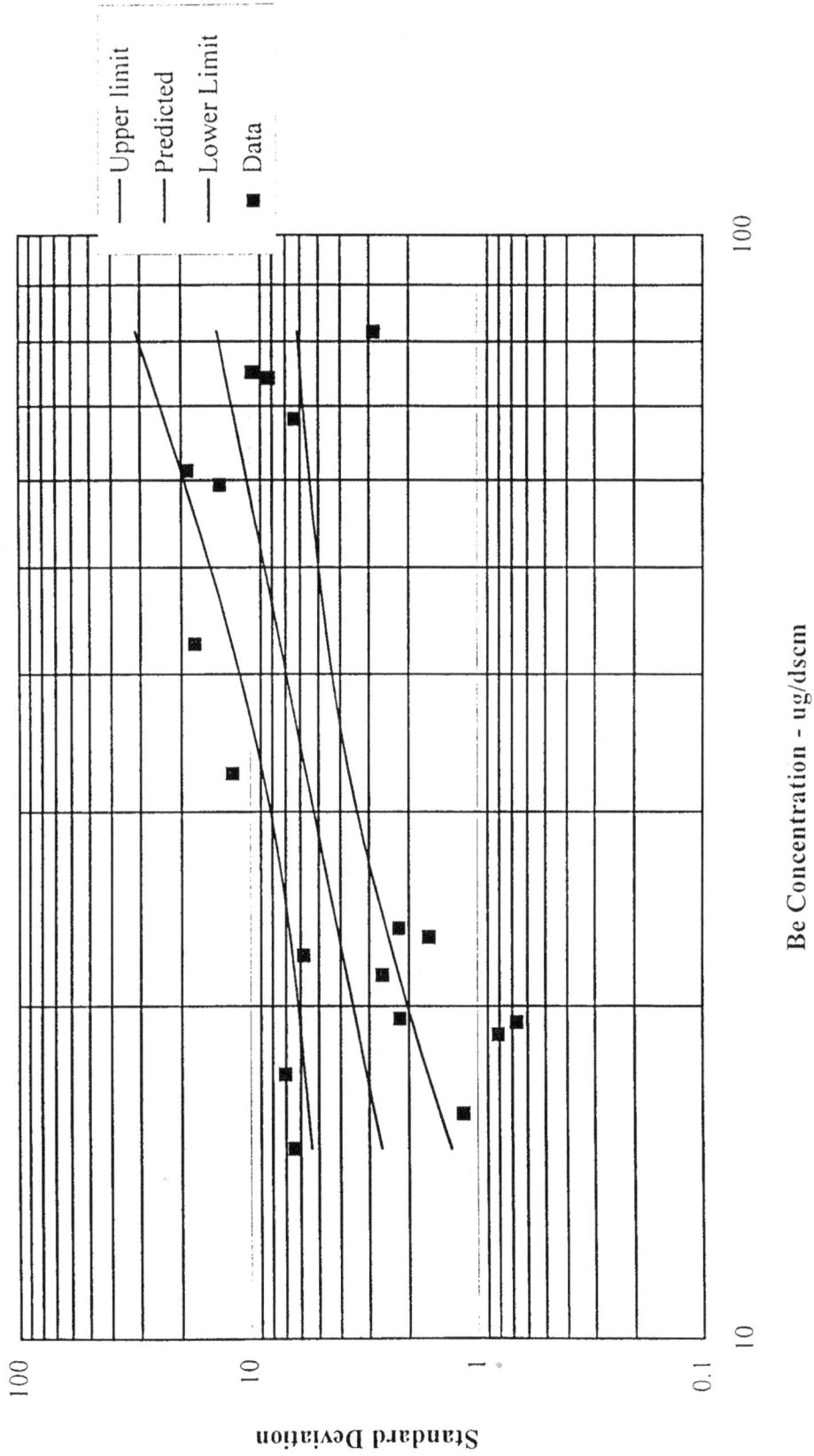

Figure 54. Regression Line and 95% Confidence Interval for
EPA Method 29 - Beryllium

144

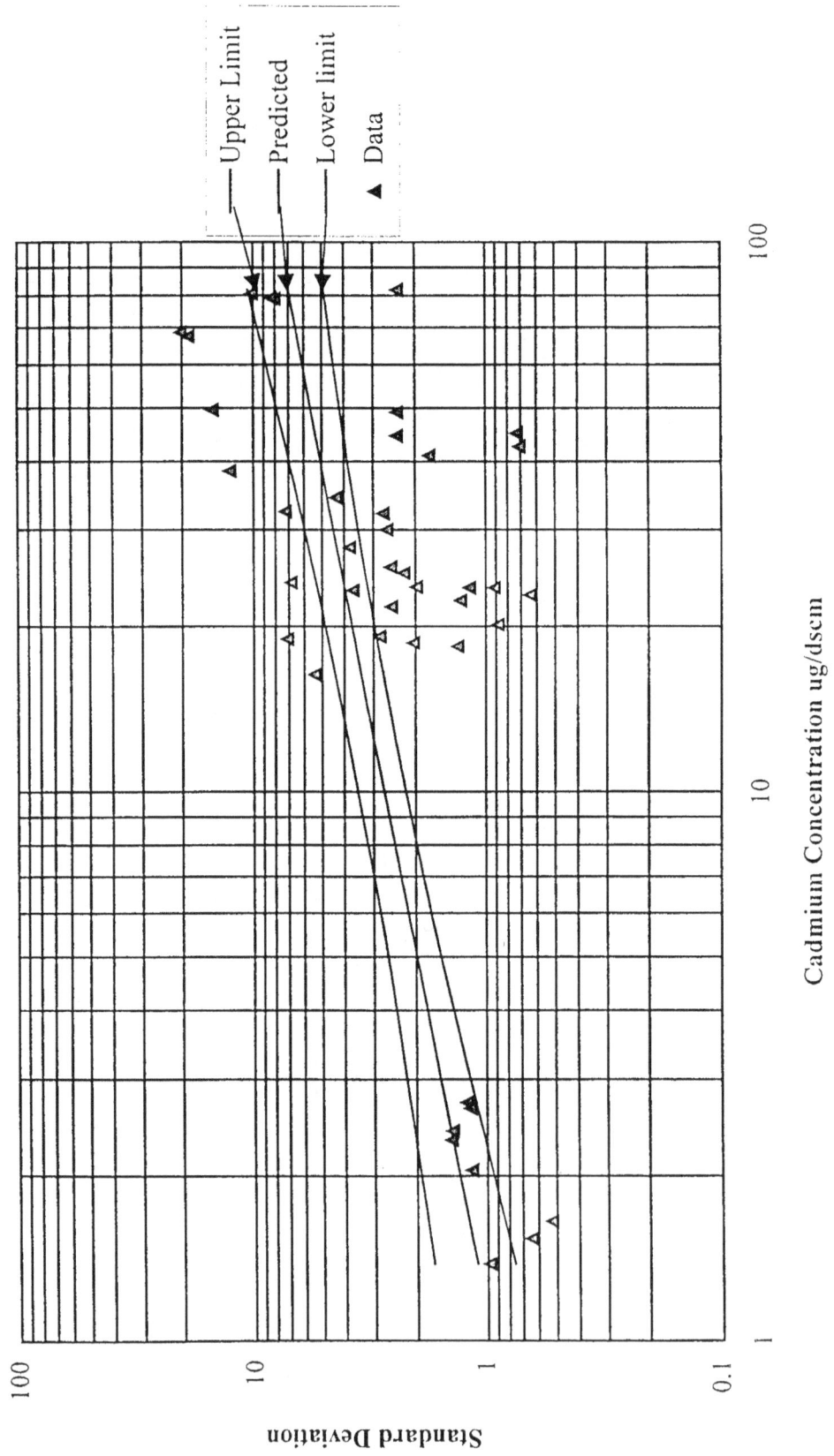

Figure 55. Regression Line and 95% Confidence Intrerval for EPA Method 29 - Cadmium

145

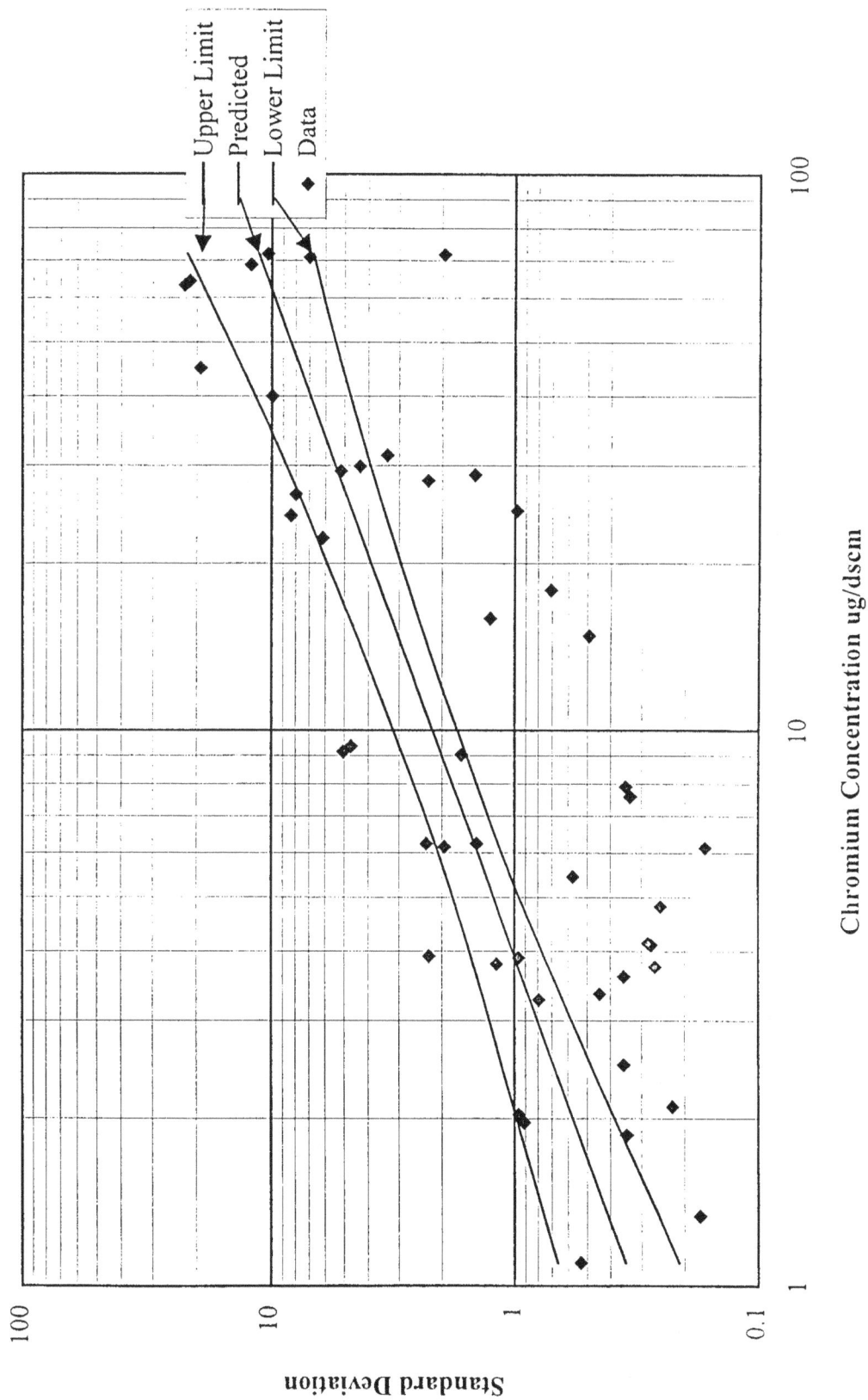

Figure 56. Regression Line and 95% Confidence Interval for
EPA Method 29 - Chromium

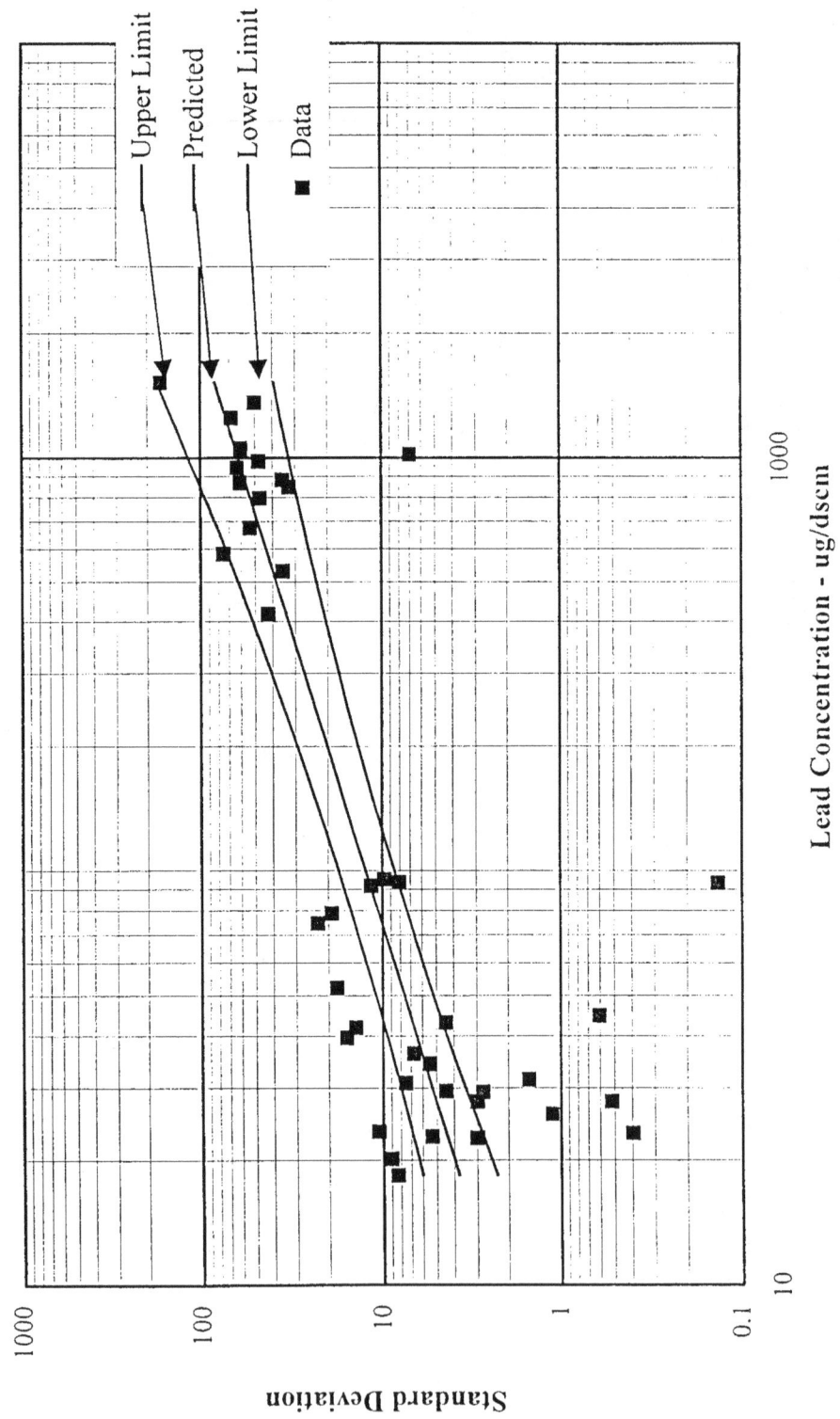

Figure 57.   Regression Line and 95% Confidence Interval for
EPA Method 29 - Lead

Figure 58. EPA Method 29 Precision Metrics – Antimony

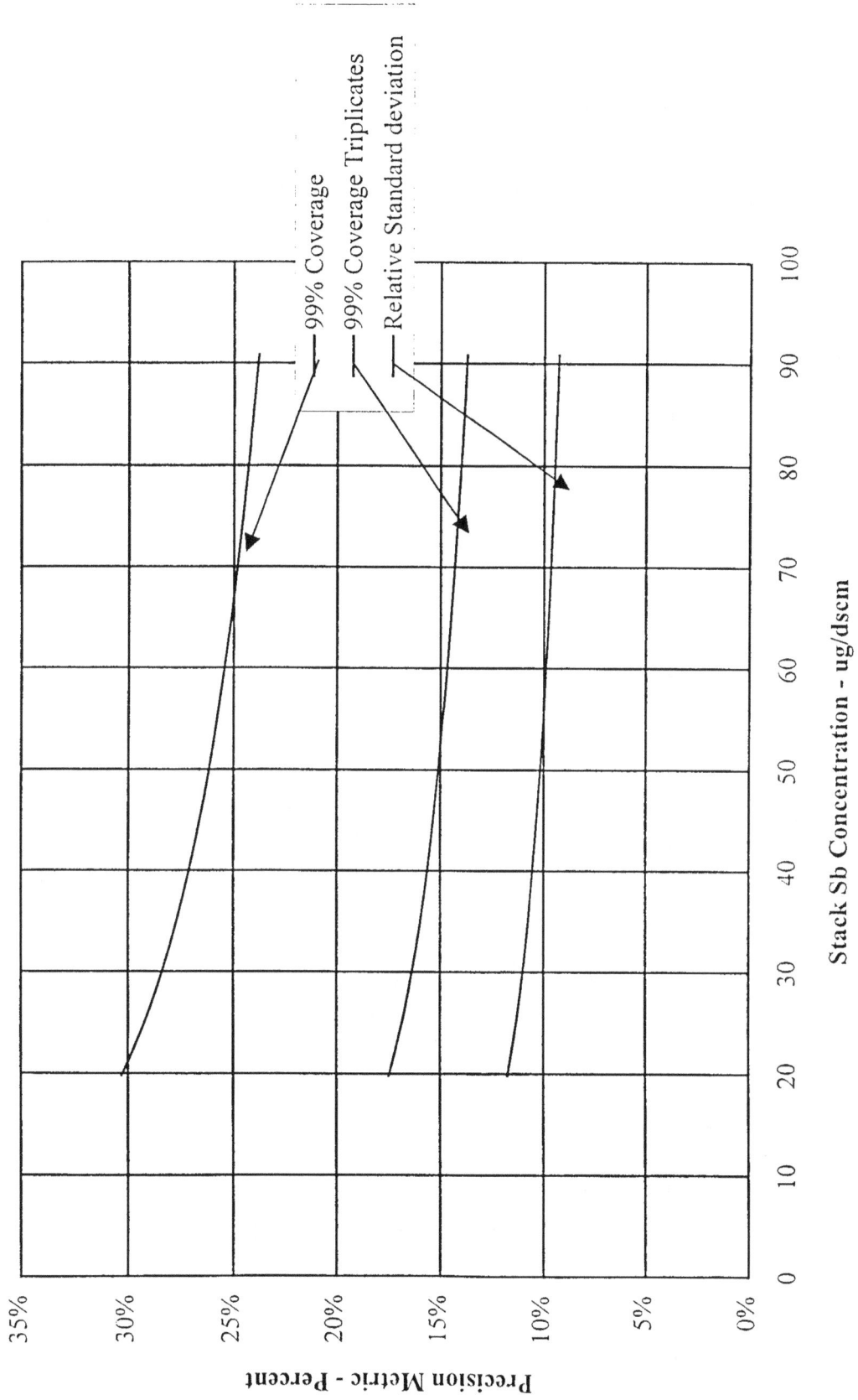

148

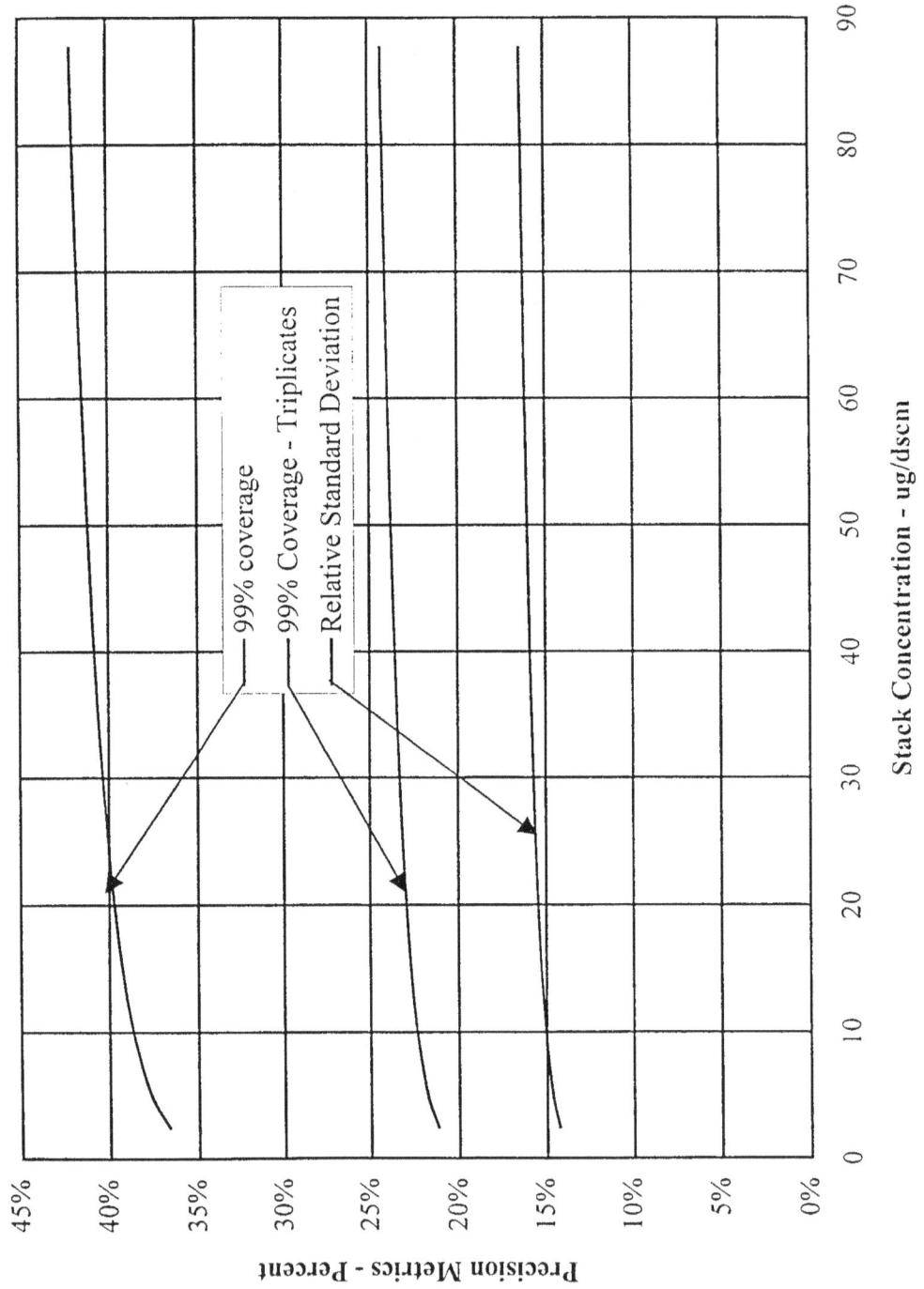

Figure 59. EPA Method 29 Precision Metrics - Arsenic

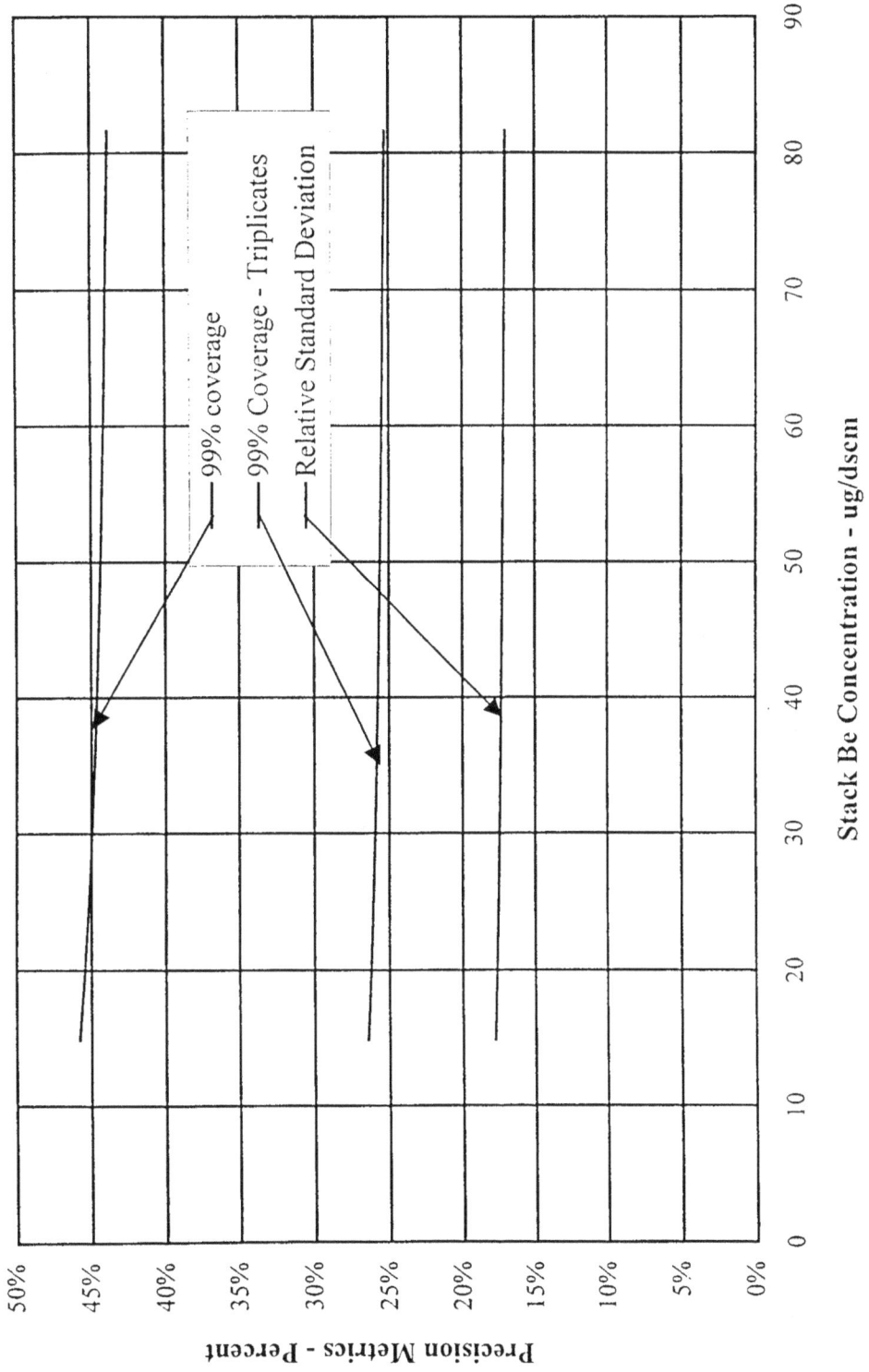

Figure 60. EPA Method 29 Precision Metrics – Beryllium

150

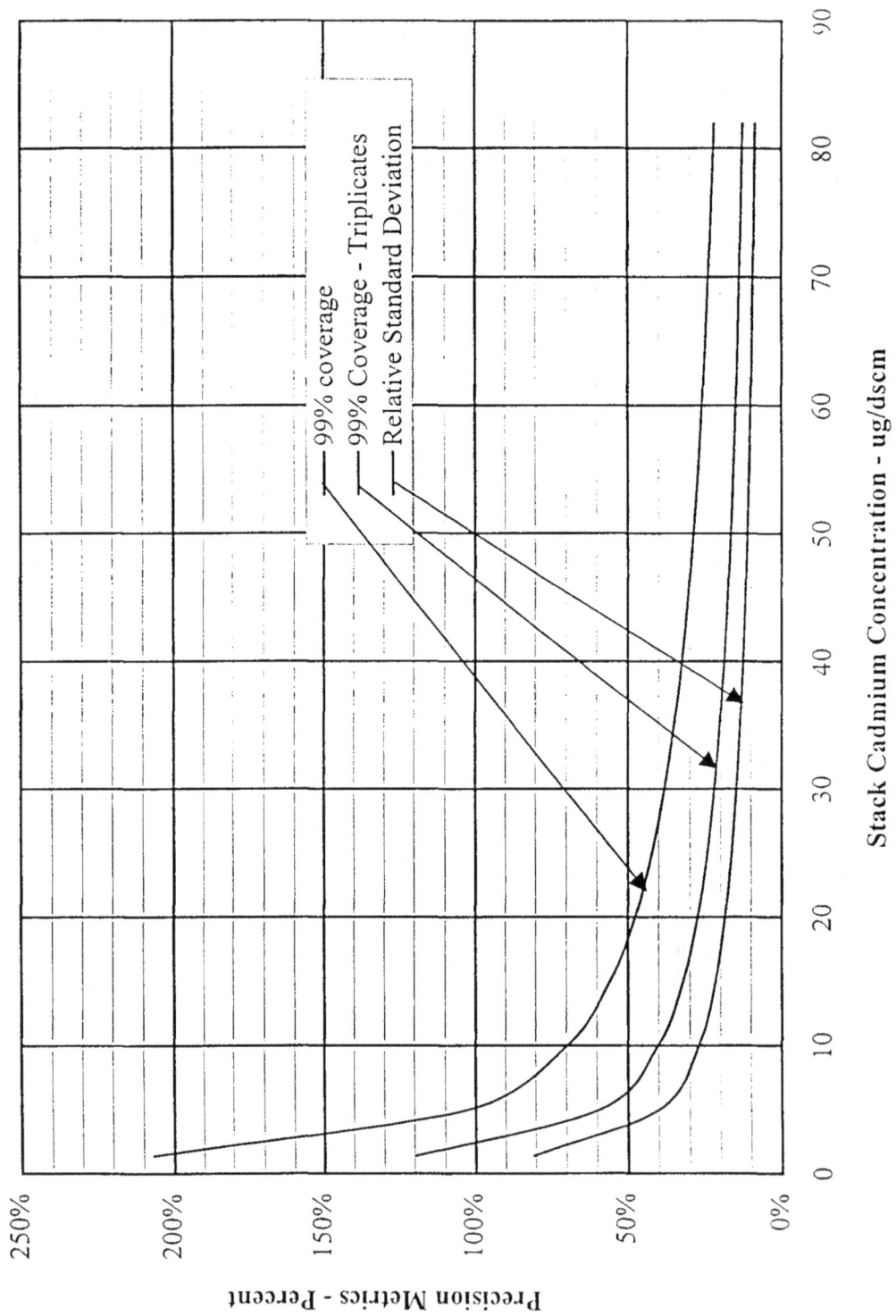

Figure 61. EPA Method 29 Precision Metrics - Cadmium

151

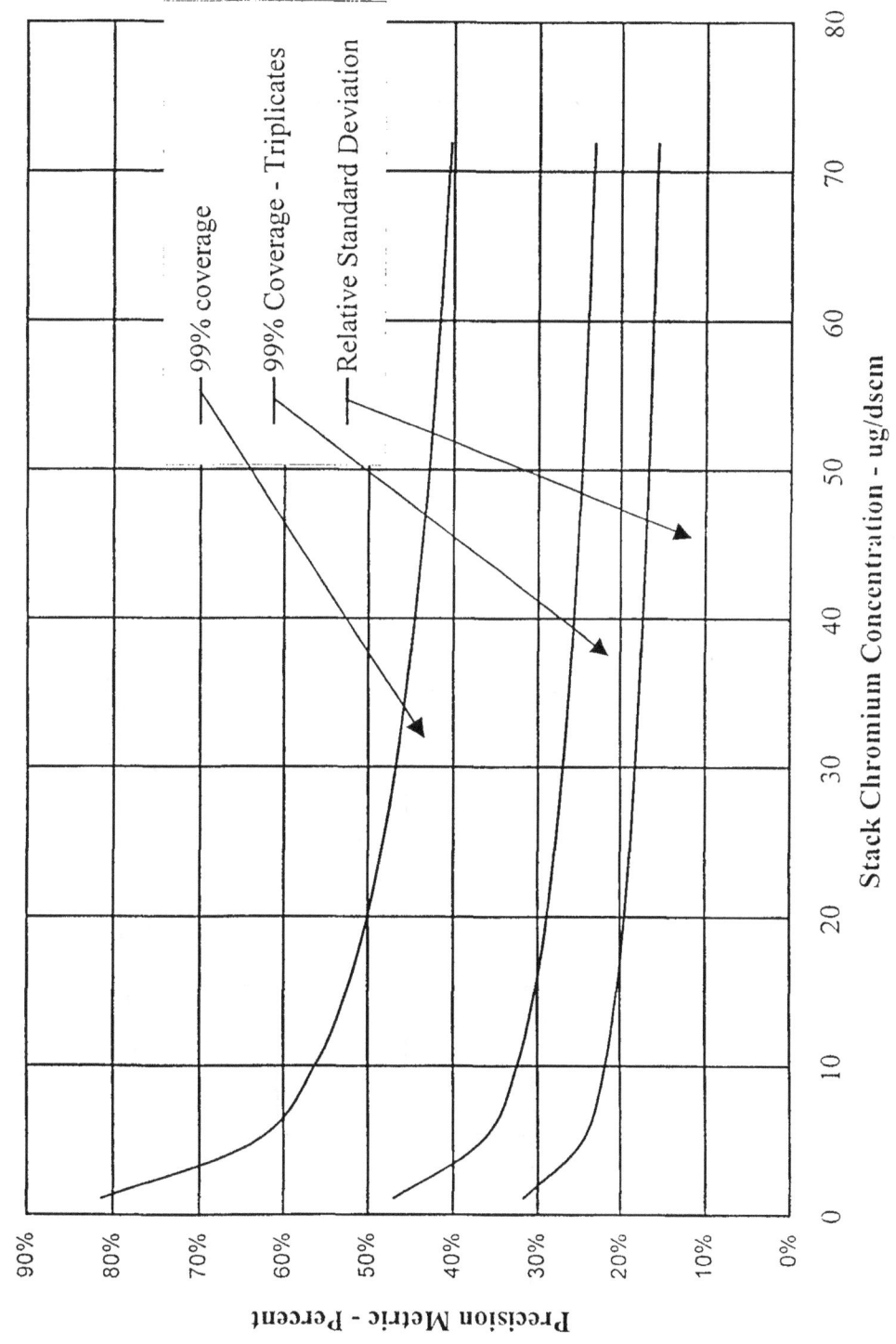

Figure 62. EPA Method 29 Precision Metrics - Chromium

152

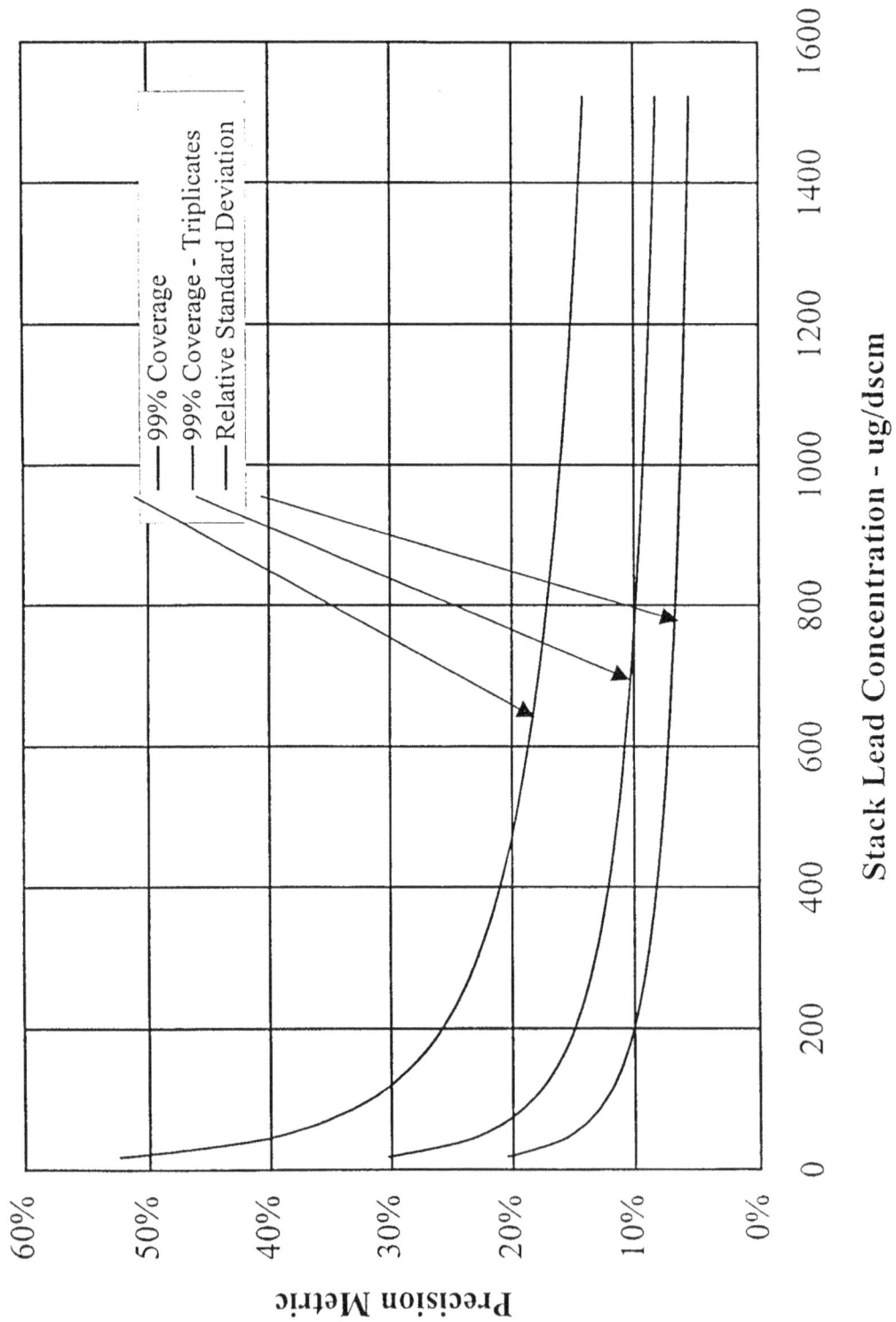

Figure 63. EPA Method 29 Precision Metrics - Lead

153

Figures 61, 62 and 63 for cadmium, chromium and lead all indicate that the relative standard deviation begins to increase rapidly as the stack concentration fall below about 10 to 20 μg/dscm. For cadmium, RSD is predicted to decrease from 18.7% at 20 μg/dscm to 9.1% at 80 μg/dscm. However, at 5 μg/dscm, the predicted value of RSD is 38.6% and at 1.4 μg/dscm, RSD is predicted to exceed 75%. For chromium and lead, less dramatic, but significant increases in RSD are anticipated when concentration drops below 20 μg/dscm.

As noted earlier, the data in Figures 58 through 63 are based on the assumption that standard deviation varies according to the regression equations. However, within 95% confidence bounds, it is distinctly possible that standard deviation could be greater or less than predicted by the regression lines. Figures 64 through 69 present the anticipated ranges for 99 out of 100 future measurements as a function of the true stack concentration, assuming that that standard deviation varies according to the upper 95% confidence limit, according to the regression analysis, and according to the lower 95% confidence limit. These figures illustrate the critical importance of the uncertainty in the overall precision assessment. As indicated by results presented in Figures 58 and 64, if the standard deviation of Method 29 antimony measurements varies according to the regression equation, then 99 out of 100 future measurements (at a stack concentration of approximately 90 μg/dscm) should fall within ± 23% of the true stack concentration. However, at the 95% confidence level, it is possible that future data might have a spread as large as ± 51.9% about the true concentration of 90 μg/dscm.

The significance of the uncertainty in the standard deviation versus concentration relation is even more dramatic for Method 29 measurements of beryllium. As indicated in Figures 60 and 66, when Method 29 beryllium measurements are made at true stack concentrations of 80 μg/dscm, available data indicate that 99 out of 100 future measurements will fall within ± 43.8% of the true concentration (45 to 115 μg/dscm). Also, 99 out of 100 triplicate measurements are expected to fall within ± 25.3% of the true stack concentration (when true concentration equals 80 μg/dscm). However, at the 95% confidence level, it is only possible to conclude that the future single measurements will fall within ± 97.5% of the true concentration. This large spread between our best estimate and the 95% confidence limits is a direct result of the limited amount of multi-train data for beryllium using Method 29. The potential range for future measurements for other metals is illustrated on the respective figures.

Figure 64. Precision Estimates For Measurements Using EPA Method 29 For Antimony

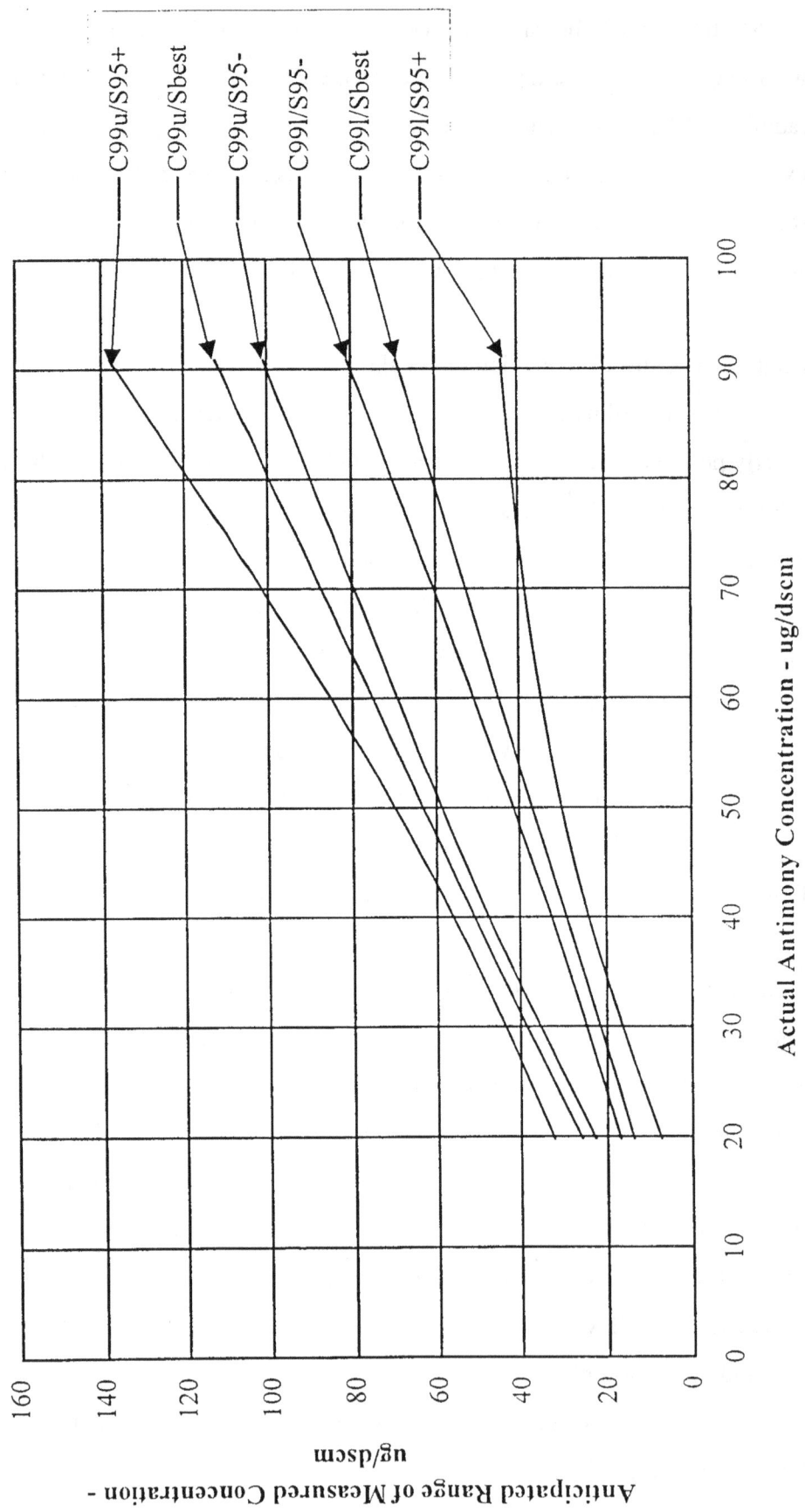

C99u/S95+
C99u/Sbest
C99u/S95-
C99l/S95-
C99l/Sbest
C99l/S95+

Actual Antimony Concentration - ug/dscm

Anticipated Range of Measured Concentration - ug/dscm

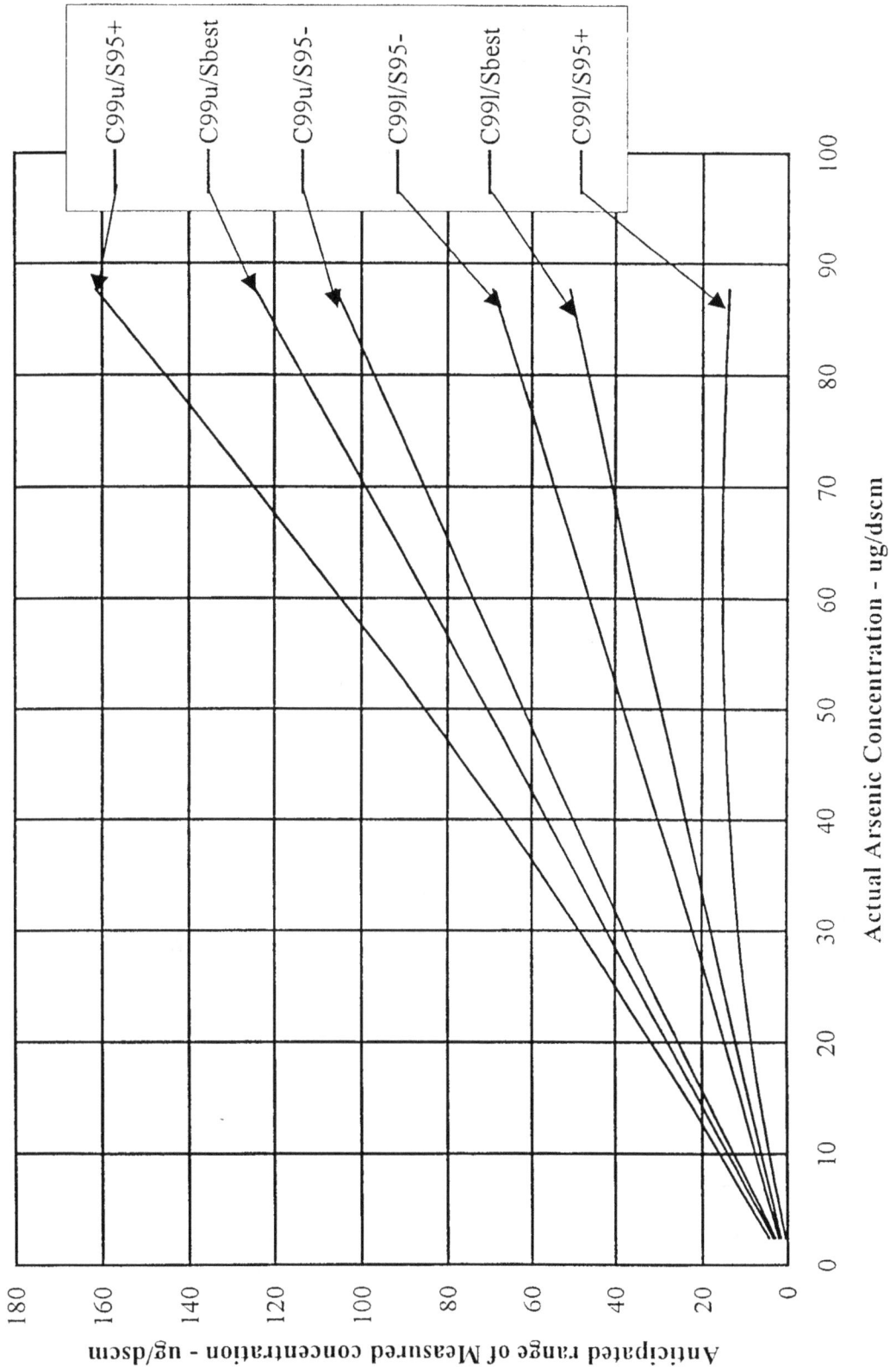

Figure 65. Precision Estimates for Measurements using EPA Method 29 for Arsenic

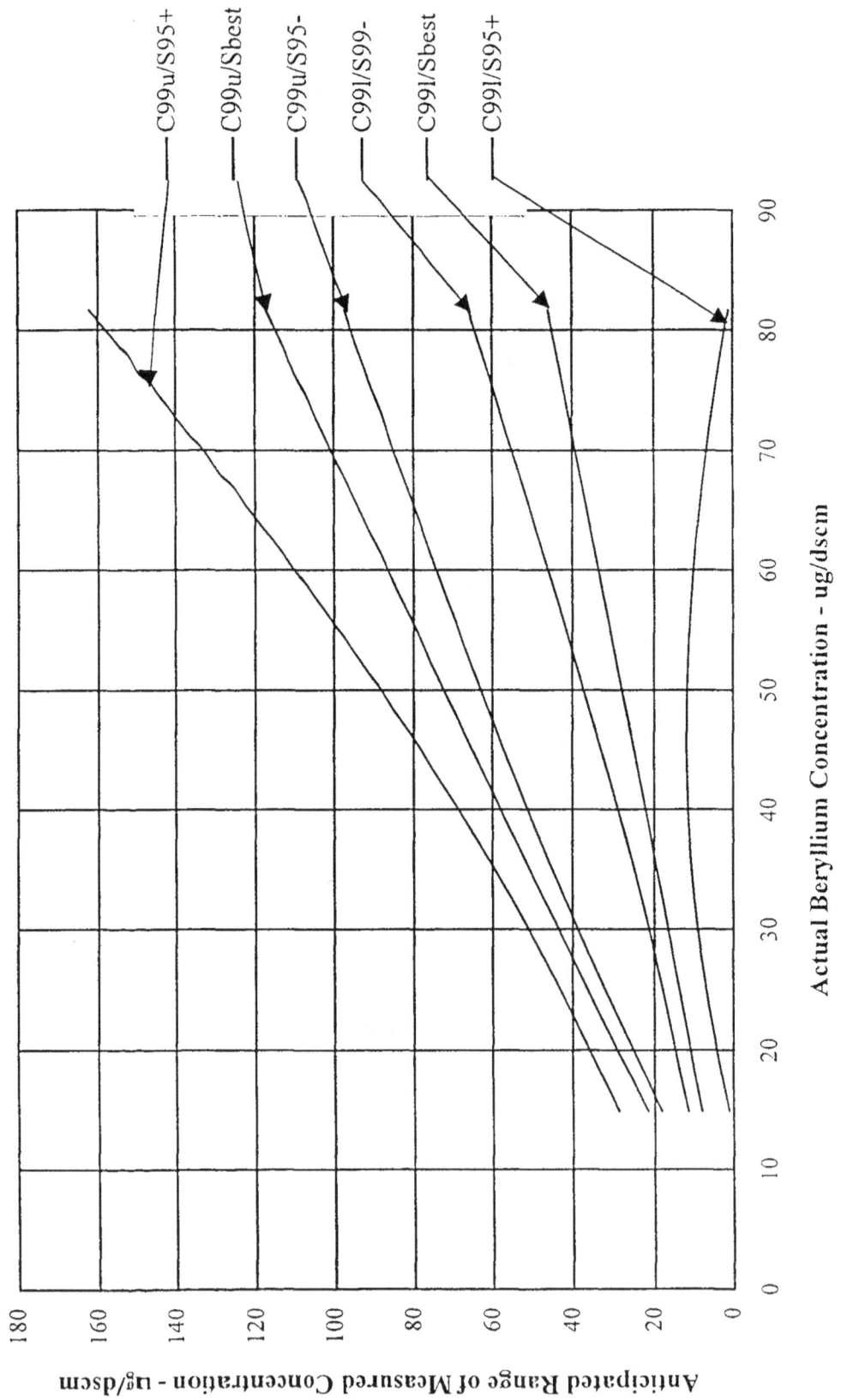

Figure 66. Precision Estimates For Measurements Using
EPA Method 29 For Beryllium

Figure 67. Precision Estimates for Measurements Using EPA Method 29 for Cadmium

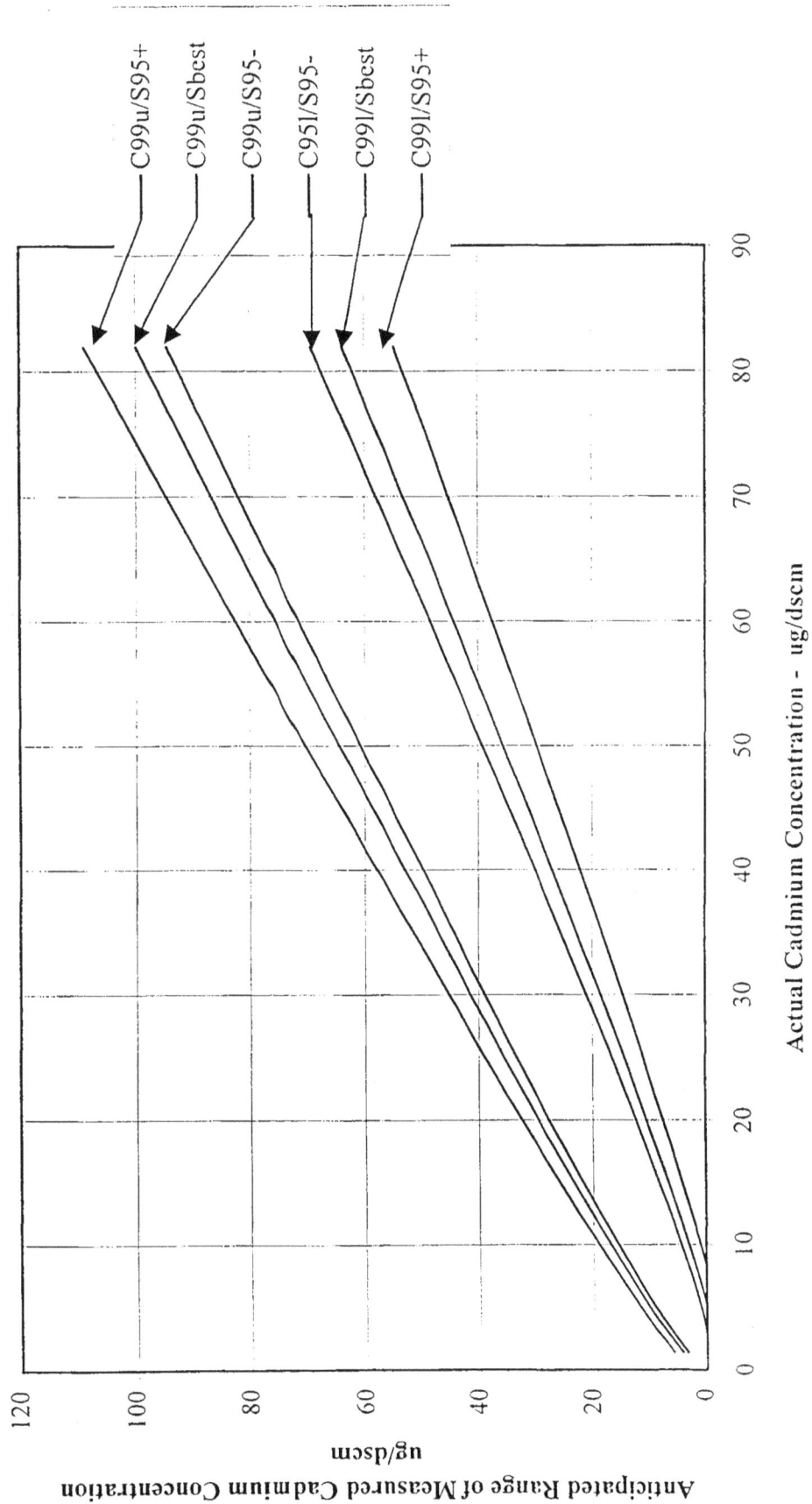

C99u/S95+
C99u/Sbest
C99u/S95-
C95l/S95-
C99l/Sbest
C99l/S95+

Anticipated Range of Measured Cadmium Concentration
ug/dscm

Actual Cadmium Concentration - ug/dscm

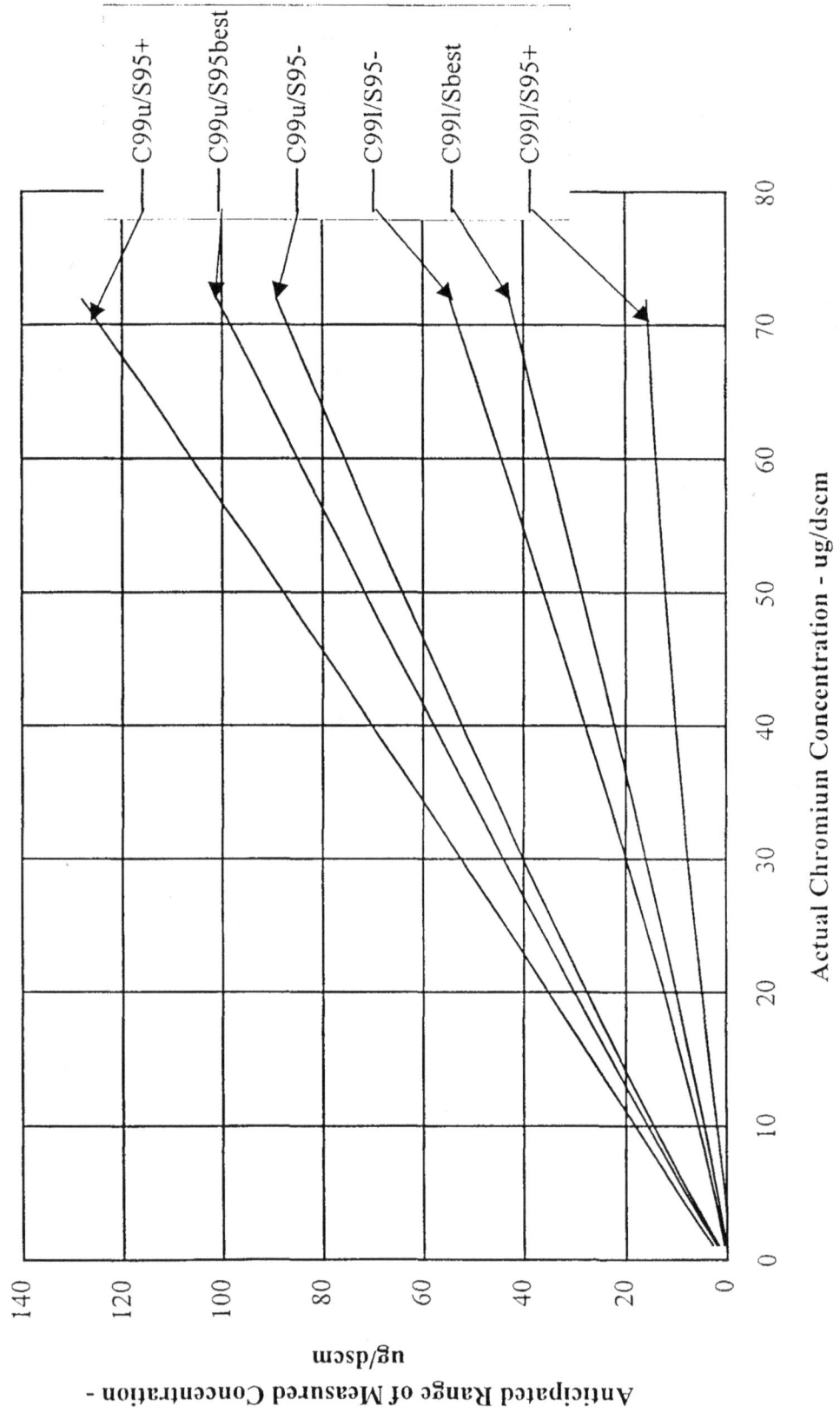

Figure 68. Precision Estimates for Measurements Using EPA Method 29 for Chromium

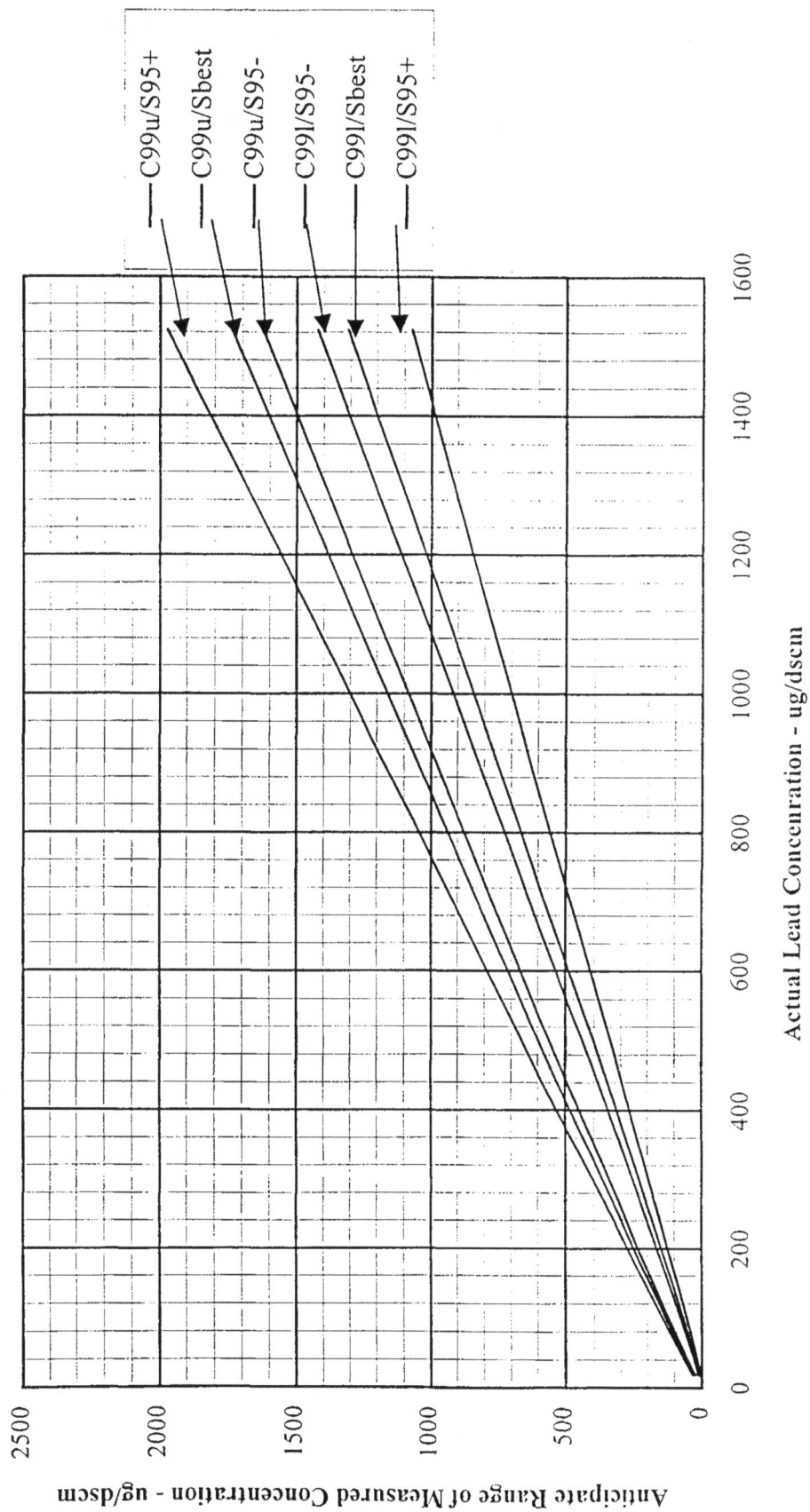

Figure 69.   Precision Estimates for Measurements Using
EPA Method 29 for Lead

Earlier discussion of the results presented in Table 24 noted a similarity in the regression analysis for five of the six metals, including antimony, arsenic, beryllium, chromium, and lead. It was pointed out that Method 29 collects a sample that is subsequently analyzed for all of the target metals. Random errors in the sample collection and sample recovery operations are likely to contribute similar random error to each of the measured metal concentrations. Finally, it was noted that random error in the sample collection and recovery process tend to drive the value of the power function coefficient toward 1.0 while random error in the analytical analysis are more typically characterized by constant standard deviation (p equals zero in the regression equation). In Table 24, note that with the exception of cadmium, the regression analysis for five metals indicates that the p coefficient is between 0.703 and 1.039 suggesting that random error in sampling and recovery are major contributors to overall measurement imprecision.

The general similarity in these regression results further suggests that for these five metals, it may be appropriate to assess a composite precision estimate for Method 29. To perform that assessment, the multi-train data presented in Tables 18, 19, 20, 22, and 23 were combined into a single data set. With the exception of the multi-train data for lead, the majority of the individual data sets were from tests where the metal concentrations ranged from single digit to less than 100 µg/dscm. Further the majority of the lead data are from this same concentration range. To form a composite data set for regression analysis, the Rigo and Chandler data were eliminated from the lead data yielding data for all five metals in the same concentration range. The data were appropriately weighted for the number of degrees of freedom and then subjected to a regression analysis. The resultant regression equation has 158 degrees of freedom and produced a t-statistic equal to 10.66 – well above the critical t-statistic. The analysis suggests that the relationship between S and concentration is described by the equation:

$$S = 0.30\, C^{0.821}$$

The value of the p coefficient (0.821) is easily within the 95% confidence ranges for p determined by regression analysis for each of the five individual metals (see Table 24). Figure 70 is a plot of the regression equations for the composite data set including the 95% confidence intervals on that

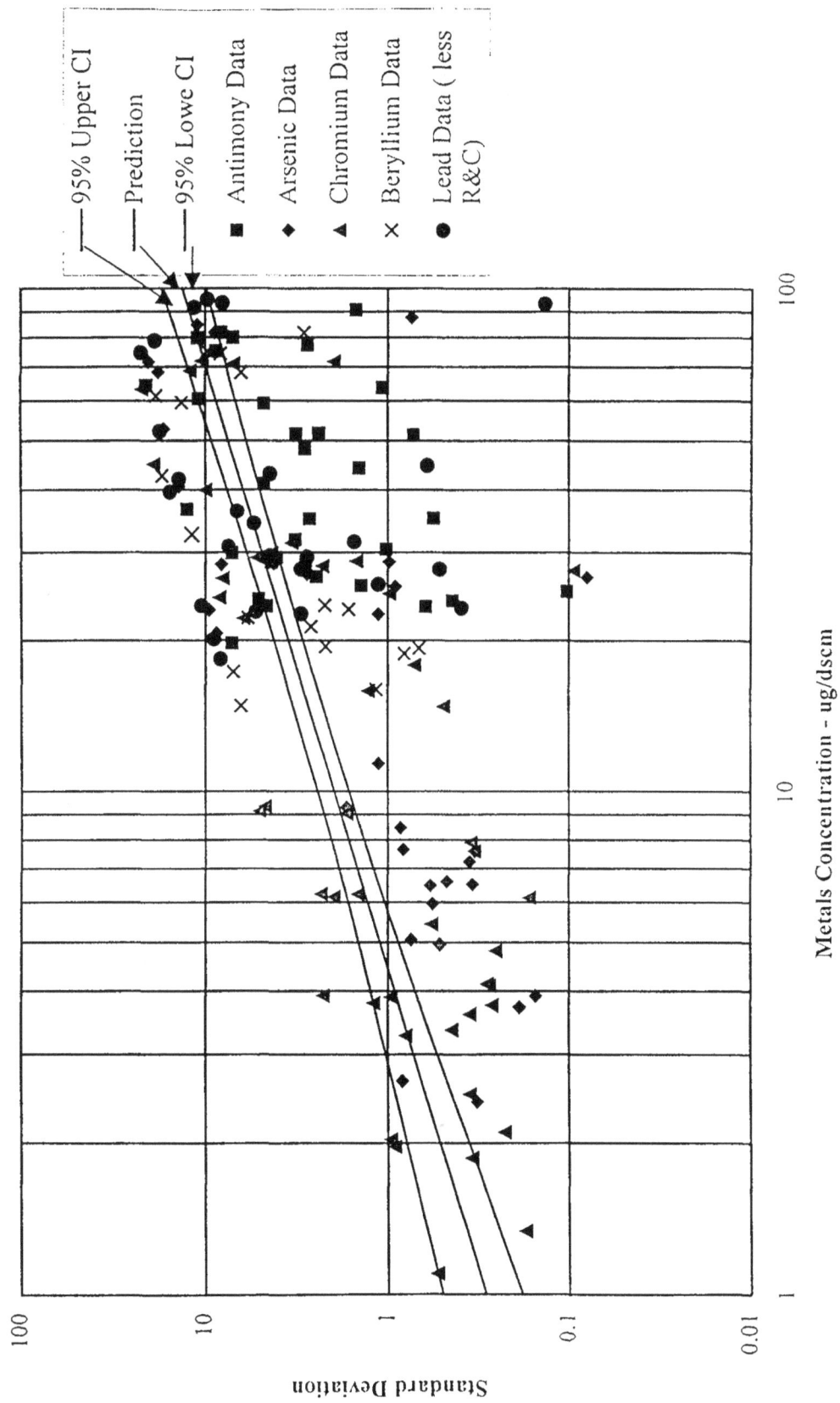

Figure 70. M29 Composite Regression Analysis and 95% Confidence Interval

162

regression. Figure 71 presents a comparison of the regression equation for each of the five individual metals and the regression equation from the composite analysis.

It is important to reiterate that only a limited body of data is available to assess the precision of Method 29. Moreover, one of the critically important data sets, the EPA pilot scale tests, contains a known bias. Including these biased results in the analysis clearly results in a slight over estimate of the standard deviation at any given stack concentration (particularly for the regression equation.) However, if those data were eliminated from the assessment, the result would be an even larger over estimate of the Method's imprecision. As a result of the limited quantity of data and the known biases in some of those data, the precision of Method 29 for metals can only be generally estimated.

The analysis of the composite data set provides a basis for the overall assessment of the method's precision. Figure 72 presents various precision metrics determined from the regression analysis of the composite data set. Based on the available data, it appears that Method 29 provides an RSD between 13 and 18% when the metal loading is between 20 and about 100μg/dscm. At metal concentrations below about 10 μg/dscm, the imprecision of the method appears to increase asymptotically. Relative to future measurements, if the precision of Method 29 conforms to the composite analysis and if the metals concentration is greater than 20 μg/dscm, 99 out of 100 single measurements should deviate from the true concentration by no more than 45%. Similarly, in the same rage, 99 out of 100 triplicate measurements should deviate from the true concentration by less than 26%. Unfortunately, the available data do not support a more definite assessment.

Table 25 provides a summary of the anticipated range of measurement results for application of Method 29 for determination of the concentration of antimony, arsenic, beryllium, chromium, and lead. Data in this table are derived from analysis of the composite data set and cover the concentration range of 4 to 100 μg/dscm. Note that the data in the table do not include correction for oxygen content in the stack.

Figure 71. M29 Composite Regresion Analysis

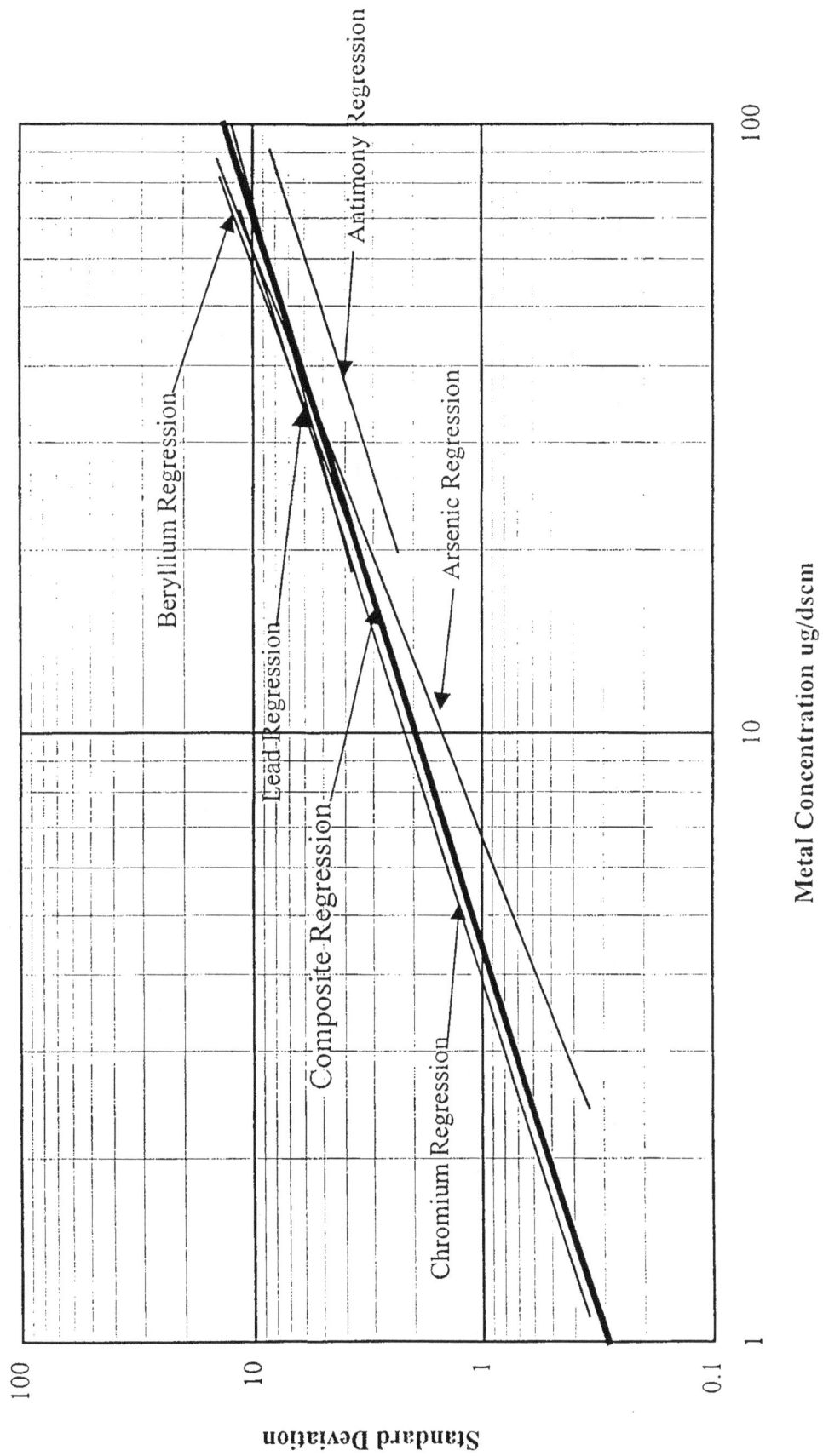

Metal Concentration ug/dscm

Standard Deviation

Beryllium Regression

Antimony Regression

Lead Regression

Composite Regression

Arsenic Regression

Chromium Regression

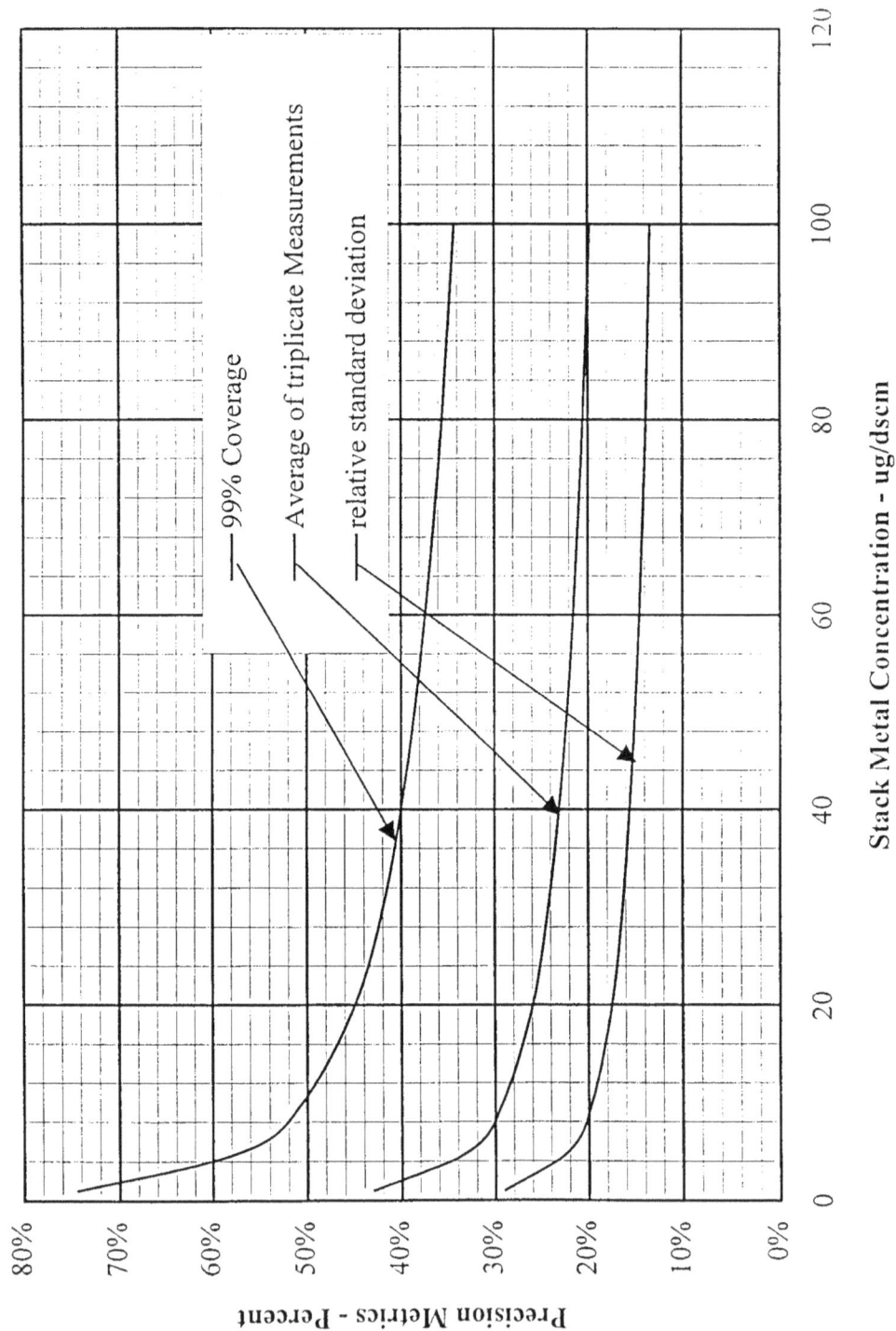

Figure 72. EPA Method 29 Precision Metrics- Composite Data

99% Coverage

Average of triplicate Measurements

relative standard deviation

Stack Metal Concentration - ug/dscm

Precision Metrics- Percent

165

Table 25.   Range of Anticipated Future Metals Data
Antimony, Arsenic, Beryllium, Chromium, or Lead

True Stack Concentration for Sb, As,Be,Cr, or Pb mg/dscm	99 out of 100 Single Measurements		99 out of 100 Triplicate Measurements	
	Lower limit	Upper Limit	Lower limit	Upper Limit
1	0.26	1.74	0.57	1.43
5	2.16	7.84	3.36	6.64
10	4.95	15.05	7.08	12.92
20	11.01	28.99	14.81	25.19
30	17.41	42.59	22.73	37.27
40	24.00	56.00	30.76	49.24
50	30.74	69.26	38.88	61.12
60	37.59	82.41	47.06	72.94
70	44.52	95.48	55.29	84.71
80	51.53	108.47	63.56	96.44
90	58.60	121.40	71.87	108.13
100	65.72	134.28	80.21	119.79

9.0 Other Measurement Methods.

An attempt was made to gather multi-train data for a variety of additional EPA measurement methods. No multi-train data or method validation test reports were uncovered for EPA methods 0030 or 0011 for volatile and semi-volatile organic emissions. The same can be said for EPA Method 23a, which is a special procedure for measuring dioxin and furan emissions. In the absence of multi-train measurement data, it is not possible to fully assess the precision of these methods.

This page Intentionally left blank.

10.   Conclusions

The ReMAP study has assembled a database containing all known multi-train data sets using various EPA measurement Methods. Data in that database have been subjected to a detailed analysis to assess the precision of the Methods as a function of stack concentration. The scope of the available data is extremely limited, especially considering the importance of results from application of the Methods. Certain of the Methods, especially Methods 5 and 5i have significant databases and the precision of those methods is relatively well established.   Using relative standard deviation as the precision metric, the precision of Method 5 is between 5 and 11% when applied to stacks with a broad range of concentrations. If applied with attention to detail, this method is capable of providing reasonably precise results, even at stack particulate concentrations as low as 15 mg/dscm. Whether an RSD of 5 to 11% is sufficiently precise is likely to be application specific.

Method 5i was specifically developed to provide precise measurement results for particulate matter concentrations below 50 mg/dscm. Based on available data, this precision of this Method has no statistically significant variation with stack concentration. Pooled analysis of the data indicates that the Method (when applied at concentrations between about 5 and 50 mg/dscm) has a characteristic standard deviation of 1.43 mg/dscm.

Method 23 for dioxin and furan emissions is a critically important method for current EPA emission regulations and for public perception of risk associated with emissions from combustion facilities. There is only a very small database for assessment of the precision of this Method. Based on available data, the anticipated RSD for measurements of the total mass of tetra through octa chlorinated dioxin and furan is estimated to be between 6.3 % and 20% for stack concentrations in the range of 2 to 27 ng/dscm (higher RSD at lower concentration). Recall that the stated range of stack concentrations are given on an "as-measured basis" and therefore do not include excess air dilution correction factors (e.g., correction to 7% O_2). The anticipated range for 99% of future individual measurements is ± 2.57 times the standard deviation.

Method 23 is also used to determine dioxin and furan emissions calculated as ITEQ. Analysis of available data for emissions expressed in this manner did not yield an acceptable regression expression. That is, the t-statistic for the regression was less than the critical t-statistic at the 95% confidence level. This implies that strong weighting of data associated with application of Toxic Equivalence Factors may be masking key trends in the data precision. By applying the TEF weighting factors, variation between simultaneous data for certain specific congeners will be significantly amplified relative to differences for other congeners. However, based on the available data and when used to determine ITEQ emission concentration, there is no statistically significant variation of Method 23 precision with concentration. A pooled analysis of the available data was performed to determine a characteristic standard deviation of the Method. That analysis indicates the pooled standard deviation of the Method is 0.027 ng ITEQ/dscm, when applied in the concentration range 0.02 to 0.9 ng ITEQ/dscm. It is critically important to note that, at the 95% confidence limit, the possible range of a single measurement is ±0.095 ng ITEQ/dscm. Thus, if a Method 23 data point indicates stack concentration of 0.095 ng ITEQ/dscm, that data point (at the 95% confidence level) could be as low as zero or as high as 0.19 ng ITEQ/dscm. Moreover, using the limited quantity of currently available data and at the 95% confidence level, it is not possible to determine compliance with a PCDD/PCDF emission limit below 0.095 ng ITEQ/dscm.

Method 26 for hydrochloric acid measurements was found to be as precise as any of the manual measurement methods. Typically, the RSD for this method was found to range between 5 and 10%. RSD does increase at very low HCl concentrations but the regression analysis suggests that RSD should only increase to 15.9% at concentrations as low as 1 mg/dscm.

Mercury emission measurements are also extremely important relative to current emission regulations. A relatively large array of multi-train emission data was available for the ReMAP analysis. The analysis of results found that the RSD for total mercury measurements by Methods 29 and 101 had minimal variation with concentration. Over the concentration range of 50 to 783 μg/dscm the measurement method RSD varied from 9.6 to 12.4 percent. A significant portion of the overall database is for measurements at relatively low mercury concentrations. Forty out of 73 mercury data points had average concentrations below 50 μg/dscm while 28 of the 73 measurements were at average stack concentration less than 30 μg/dscm. As concentration drops from 50 to 5 μg/dscm, the ReMAP analysis shows that RSD is anticipated to increase from 12.4 to 15.4%.

Method 29 is also used for measurement of other metals. Precision analysis was completed for six other metals including antimony, arsenic, beryllium, cadmium, chromium, and lead. The analysis shows that five of the six metals all behave similarly with respect to measurement method precision. Data for all metals except cadmium exhibit a standard deviation versus concentration relation where the power function coefficient (p) has a value of approximately 0.82. Composite analysis of data for this group of metals suggests that Method 29 provides an RSD on the order of 13 to 18% when the metal loading is between 20 and about 100µg/dscm. Data for lead is available at much higher concentration and the method RSD for lead appears to asymptote between 5 and 10%. At metal concentrations below about 20 µg/dscm, the imprecision of the Method appears to increase asymptotically. For cadmium, based on the available data standard deviation of the Method has a different relation with concentration. The indicated value of the p coefficient is approximately 0.45 suggesting that for the available cadmium data, random error (or differences) in the chemical analysis was a significant contributor to the overall imprecision of the Method.

One additional conclusion from the ReMAP Phase 1 project is that there is a pressing need for additional multi-train data to refine the precision estimates of the EPA Reference Methods. If such experimental programs are to be conducted, significant attention should be given to the appropriate range of stack concentrations. Results presented in this report can help to guide the test planning efforts.

REFERENCES

1. American Society of Mechanical Engineers, <u>Test Uncertainty, Instruments and Apparatus,</u> ASME Report ASME PTC 19.1-1998.

2. Midgett, M.R. "How EPA Validates NSPS Methodology." Environmental Science and Technology. 11(7): 655-659, 1977.

3. EPA Reference Method 301--Field Validation of Pollutant Measurement Methods from Various Waste Media. 57 FR 61970, December 29, 1992.

4. EPA Reference Method 5 – Determination of Particulate Emissions from Stationary Sources. 52 FR 22888, June 16, 1987 (also includes various revisions).

5. Collaborative Study of Particulate Emissions Measurements by EPA Methods 2, 3, and 5 using Paired Particulate Sampling Trains (Municipal Incinerators). H. Hamil and R. Thomas, Southwest Research Institute, EPA Contract No. 68-02-0626, EPA-600/4-76-014. March 1976.

6. NESHAPS: Standards for Hazardous Air Pollutants from Hazardous Waste Combustors; Final Rule. 62 FR 52828 dated September 30, 1999.

7. Particulate Matter CEMS Demonstration at the DuPont On-Site Incinerator in Wilmington, DE. Energy and Environmental Research Corporation for EPA Office of Solid Waste and Emergency Response. October 1997. (also see 62 FR 67788)

8. Particulate Matter CEMS Demonstration at the Eli Lilly Incinerator in Clinton, IN. Prepared by Eli Lilly in association with CMA and CRWI.

9. EPA Reference Method 23 – Determination of Polychlorinated Dibenzo-p-Dioxins and Polychlorinated Dibenzofurans from Stationary Sources. 56 FR 5758, February 13, 1991 (with several revisions).

10. Validation of Emission Test Method for PCDDs and PCDFs. Prepared by Midwest Research Institute, EPA Contract No. 68-02-4395, Work Assignment 23 for EPA EMSL. February 24, 1989.

11. Retrofit of Waste to Energy Facilities Equipped with Electrostatic Precipitators. Prepared by Rigo & Rigo Associates and A.J. Chandler and Associates under the Direction of ASME. June 1997.

12. Dioxins/Furans, HCl, Cl_2 and Related Testing at a Hazardous Waste Burning Light-Weight Aggregate Kiln. Prepared by Energy and Environmental Research Corporation, Contract No. 68-D2-0164 for the EPA Office of Solid Waste. October 10, 1997.

13. EPA Reference Method 23 – Determination of Polychlorinated Dibenzo-p-Dioxins and Polychlorinated Dibenzofurans from Stationary Sources. 56 FR 5758, February 13, 1991.

14. OMSS Field Test Report on Carbon Injection for Mercury Control Completed at the Ogden Martin of Stanislaus, Inc. Prepared by Radian Corporation. Contract No. 68-D10010 for the EPA Office of Research and Development. September 1992.

15. EPA Reference Method 29 – Determination of Metals Emissions from Stationary Sources. 61 FR 18262, April 25, 1996.

16. EPA Reference Method 101a – Determination of Mercury Emissions from Sewage Sludge Incinerators. 47 FR 24703, June 8, 1982.

17. Proposed Draft EPA Method 101b – Determination of Mercury Species from Stationary Sources.

18. Validation of Draft Method 29 at a Municipal Waste Combustor. Prepared by Radian Corporation, Contract No. 68-D9-0054 for the EPA Emission Measurements Branch. September 30, 1992.

19. Mercury CEMS Demonstration at the Holnam, Inc. Hazardous Waste Burning Cement Kiln in Holly Hill, SC. Energy and Environmental Research Corporation for EPA Office of Solid Waste and Emergency Response. October 1997. (also see 62 FR 67788).

20. Internal EPA Study at the Pilot Scale Rotary Kiln Incinerator Located in Research Triangle Park, NC. Sampling and Analysis performed by EPA.

21. Emissions of Metals and Organics from Municipal Wastewater Sludge Incinerators. Prepared by Entropy and DEECO, Contract No. 68-CO-0027, for EPA Risk Reduction Engineering Laboratory.

Appendix

Statistical Analysis Procedures
for the ReMAP Program

Prepared by:

Charlie Hendrix
Statistical Consultant

Procedures for Analyzing Simultaneously Sampled Concentration
Data to Determine Measurement Precision (Random Error)

This appendix is concerned with the statistical methods used in the analysis of simultaneously sampled data.

Confidence Interval on a Mean

Suppose we have collected the following replicate data from a process operating under fixed conditions.

$$103.2 \quad 107.9 \quad 101.6 \quad 109.1 \quad 105.3$$

The average of these is 105.42; the standard deviation is 3.132 with 4 degrees of freedom (df). If these are representative data from this process, then the 95% confidence interval on the true mean μ is $105.42 \pm t^*3.132/\sqrt{5}$ where t is the reference value of t with 4 degrees of freedom (df). $t = 2.776$. The 95% interval on μ is 105.42 ± 3.89 or 101.53 to 109.31. "95% of the intervals calculated in this manner will encompass the true mean, μ." From this we infer that "the probability is 95% that the interval 101.53 to 109.31 has captured the true mean, μ." If $N > 4$ then concerns about whether the data came from a non-normal distribution have little bearing on this calculation under most circumstances.

Confidence Interval on a Standard Deviation

The following replicate data also came from a process operating under a fixed set of conditions.

$$1.03 \quad 1.24 \quad 0.91 \quad 1.36 \quad 0.97 \quad 1.22$$

The standard deviation of these data is $S = 0.177$ with 5 degrees of freedom (df). If these are representative data from this process, then the 95% confidence interval on the true standard deviation σ is calculated in the following manner. Go to Table 1; enter with df = 5; find the factors 0.624 and 2.453 under the heading "For 95% Conf. Int." Multiply 0.177 by each of these factors to obtain 0.110 and 0.434.

Table 1

df	P₀.₀₂₅	P₀.₉₇₅
	For 95% Conf. Int.	
1	0.446	31.941
2	0.521	6.325
3	0.566	3.727
4	0.599	2.875
5	0.624	2.453
6	0.644	2.202
7	0.661	2.035
8	0.675	1.916
9	0.688	1.826
10	0.699	1.755
12	0.717	1.651
15	0.739	1.548
20	0.765	1.444
25	0.785	1.380
30	0.799	1.337
50	0.837	1.243
100	0.879	1.161
200	0.911	1.109
500	0.942	1.066

"95% of the intervals calculated in this manner will encompass the true standard deviation σ." From this we infer that "the probability is 95% that the interval 0.110 to 0.434 has captured the true standard deviation sigma (σ)". This assumes that the data came from a process with an underlying "true standard deviation σ" whose value is not known. The accuracy of confidence intervals on σ is affected by non-normality in the original data. Table 1 is derived from the Chi-squared distribution.

Confidence Interval on Sigma - Multiple Measures of Variation

Suppose we have data from six tests. Each pair of data was obtained by drawing simultaneous samples and analyzing those samples for the concentration of a pollutant.

Time	Data		Standard Deviation, S	df
2:30 PM	27.2	30.0	1.980	1
4:15 PM	19.1	23.7	3.253	1
5:05 PM	28.3	26.8	1.061	1
5:50 PM	20.9	27.2	4.455	1
6:25 PM	28.7	33.1	3.111	1
7:15 PM	25.5	24.2	0.919	1

We are concerned with the inherent variation due to the sampling and analysis procedures (measurement precision), measured as the variation *within* tests. The standard deviation of each pair of samples is shown.

Our objectives are to (1) estimate the standard deviation σ due to sampling and analysis (measurement precision) and (2) to calculate a 95% confidence interval on that estimate of σ.

It might seem logical to average the standard deviations and report that as the estimate of σ. Unfortunately that average will be a *biased* estimate of σ. This bias can be substantial. A more accurate estimate of σ is found by *pooling* the standard deviations. Pooling requires squaring the individual standard deviations to obtain variances; weight-averaging the variances to obtain the pooled variance; then take the square root to find the pooled standard deviation. Each variance is weighted by the number of degrees of freedom associated with that variance. The number of degrees of freedom associated with the pooled standard deviation is the sum of the degrees of freedom for the individual standard deviations.

$$\text{Pooled Variance} = (\text{Sum of df*Variance})/(\text{Sum of df})$$

$$= [1^*(1.980)^2 + 1^*(3.253)^2 + ... + 1^*(0.919)^2]/[1 + 1 + 1 + 1 + 1 + 1]$$

$$= 45.998/6 = 7.666 \qquad \text{Pooled S} = \sqrt{7.666} = 2.77 \text{ with 6 df.}$$

Pooled S is an estimate of σ. We can calculate the 95% confidence interval on σ by entering Table 1 with 6 df to find the factors 0.644 and 2.202. Multiply 2.77 by each of these factors to find 1.78 and 6.10. The probability is 95% that this interval has captured the true value of σ.

Can we calculate a confidence interval on σ by treating the individual standard deviations (1.980, 3.253, ... , 0.919) as "data"; calculate the standard deviation of those "data" and then calculate confidence limits on the average of those "data" as if we were calculating the confidence interval on an average? The answer is a qualified "yes", but only after taking into account some issues that have not been addressed at this point. This strategy has been used for the analysis of the ReMAP data. Although the ReMAP procedure involves fitting the data to a model by least squares and correcting for biases, the underlying principle is founded on weighted averages of standard deviations.

The ReMAP Procedure

The structure of the ReMAP data precludes direct pooling to estimate σ and confidence bounds on σ. This is because σ is not a constant, but is functionally related to concentrations of pollutants *and* the primary data is not positioned at just a few points on the scale of concentration.

As a consequence of this, the ReMAP procedure was designed to (1) establish the relationship between estimates of σ and concentration, and (2) calculate confidence bounds on that estimate in one pass. This is done by fitting individual values of S to a relationship of the form $S = kC^p$ where C is the observed average concentration of a pollutant. In practice this model is linearized to $Ln(S) = Ln(k) + p*Ln(C)$ before fitting. The least squares process minimizes the sum of squares of deviations in Ln(S) rather than in S.

Alternative routes were explored. For instance, nonlinear modeling could be used to estimate k and p by minimizing the sum of squares of deviations in S directly. One difficulty stems from the fact that σ varies with C; therefore the deviations between observed and predicted (the residuals) will be functionally related to C; the usual requirement of constant variance will not met. One resolution for this would be to use a weighted nonlinear least-square method. The ultimate complications in this route distract from one of the objectives. That is, in addition to obtaining relationships between unbiased estimates of σ and C, it is our intention that the ReMAP process can be understood and reproduced by those who are not highly-trained in complex statistical methods.

Because the ReMAP procedure involves biased estimates, there were concerns about whether the finished product meets the requirements. For this reason, the ReMAP procedure has been simulated using synthetic "data" that resembles the actual data; those "data" were analyzed using the ReMAP procedure in its entirety; and the outcomes were summarized to see whether (1) the final estimated of σ are unbiased and (2) whether the confidence intervals do indeed capture the true values of σ with the expected frequency. Tens of thousands of synthetic data were generated in this manner. Cases of N = 12 pairs of data; N = 24 pairs of data; N = 48 pairs of data; and other cases were examined. The simulated data were sometimes scattered along the scale of C; sometimes concentrated in two regions on the scale of C. These Monte Carlo simulations show that the ReMAP procedure yields unbiased estimates of σ and also show that the confidence intervals on (sigma) are approximately as stated. This will be discussed later.

The ReMAP Model S = kC^p

Out of all the possible models that could be used to associate
S with C, why was $S = kC^p$ chosen? Is this model adequate for all
of the pollutants?   These questions... and others concerning this model
form... are discussed in more detail in Fitting Standard Deviation vs.
Concentration Data to Alternative Models.

In choosing a model form, one of the first requirements is that the
variation in the residuals (residual = difference between observed and
predicted) should be constant along the scale of concentration. This is
called "homogeneous variance". Simple graphics depicting S as a function
of C show that this requirement is not met in the ReMAP data... at least
not in the instances in which those data span realistic ranges on the scale
of C. Before fitting the data to a model we need to stabilize the variance.
Historically this has been done by linearizing $S = kC^p$ to
$Ln(S) = Ln(k) + p*Ln(C)$. When the data are presented in Ln-Ln coordinates,
both the graphics and statistical tests confirm that the variance is
substantially homogeneous along the scale of concentration.

Typical values of p range from 0.5 to 0.9. The exceptions were
pollutants in which the concentration spanned narrow ranges and the
confidence intervals on p were wide. Since values of p have been associated
with sampling variation as opposed to variation due to chemical analysis,
the power-law model has merit aside from   its statistical properties.

A fundamental difficulty with the ReMAP data lies in the fact that
these data are poorly arranged along the scale of concentration. Rather than
being concentrated at two or three positions on that scale, the data are often
thinly scattered; sometimes concentrated near the centroid and sparse at the
extremes; seldom focused near the ends of the concentration scale.
Furthermore, data collected from one source are sometimes positioned near
one level of concentration and data from other sources are positioned at other
levels. In order to obtain precise estimates of p, we need data that spans
decades (factors of 10) on the scale of concentration; in reality there are
instances in which the data barely spans one decade. These factors virtually
preclude building models more complex than the form $S = kC^p$. Where there
were questions about this, the adequacy of this model form was testing by
the usual statistical procedures; in every instance it was found that a more
complex model could not be justified.

This is not to claim that a power-law model will be adequate for all
future data.   The choice of model form is presently limited by the factors
noted above.

The Relationship Between Sigma (σ) and Sample Standard Deviations

A standard deviation S calculated from data is an estimate of sigma. Sigma is the standard deviation of the source of the data. When we estimate σ from data, we want our estimate to "aim at the truth" (accuracy, unbiased); and we want our estimate to be "close to the truth" (precise; small variation around the truth.)

If we calculate S from a sample of data... and another value of S using other data from that same source... those two values of S will be different. Can we improve our estimate of σ by averaging values of S? Yes, we can. Averaging improves precision. Averaging still more sample values of S would reduce the width of the confidence interval around σ.

But there is a problem. S is a *biased* estimate of σ. Averaging values of S will improve precision, but that average will not be accurate. It will be a biased estimate of σ.

If we collect two simultaneous samples, the standard deviation of that sample must be multiplied by 1.253 to make it an unbiased estimate of σ. With larger amounts of data in the sample, the bias is smaller. Here is a brief table of bias correction factors.

N	Bias Correction Factor
2	1.253
3	1.128
4	1.085
5	1.064
10	1.028

Remember that each value of S is a biased estimate of σ. Averaging values of S reduces the *variation* in our estimate of σ; this average will be more precise than single values. But that average will still not be accurate; it will be a biased estimate of σ.

If we intend to use the average of S as an estimate of σ, then we must multiply each value of S by its appropriate bias correction factor before averaging. *The bias correction factor is not related to the number of sample standard deviations we accumulated to get the average standard deviation; it is related to the sample size used to to get the individual sample standard deviations.*

An Example

Table 2 shows data in which the standard deviation varies with the average. Our purpose is to find an empirical relationship between S and the sample averages. For N = 2, unbiased S = 1.253*S.

Table 2

Sample	Data		Avg, C	S	Unbiased S
1	893.7	1080.2	986.95	131.88	165.25
2	2240.4	2127.0	2183.70	80.19	100.48
3	2070.3	2219.7	2145.00	105.64	132.37
4	529.4	553.3	541.35	16.90	21.18
5	2351.1	2652.2	2501.65	212.91	266.78
6	358.9	342.7	350.80	11.46	14.36
7	1463.0	1169.4	1316.20	207.61	260.14
8	2549.1	2320.8	2434.95	161.43	202.27
9	2488.9	2470.3	2479.60	13.15	16.48
10	1415.8	1453.8	1434.80	26.87	33.67
11	743.6	655.9	699.75	62.01	77.70
12	1356.8	1392.8	1374.80	25.46	31.90
13	960.3	1099.7	1030.00	98.57	123.51
14	1949.0	1803.1	1876.05	103.17	129.27
15	247.2	297.6	272.40	35.64	44.66
16	1866.4	2247.1	2056.75	269.20	337.31

Fig. 1 shows the relationship between S and Avg. C. Fig. 2 presents the same data in Log-Log coordinates. The nature of this relationship is more visible in Fig. 2 than in Fig. 1. These data will be fit to the model $S = kC^p$ after linearizing to $Ln(S) = Ln(k) + p*Ln(C)$. This is equivalent to working in Log-Log coordinates as in Fig. 2.

By fitting the data in Table 2 to $Ln(S) = Ln(k) + p*Ln(C)$ we not only obtain estimates of k and p but also certain statistics that tell us how accurately this model will predict σ. A least squares analysis of these data is shown in Table 3.

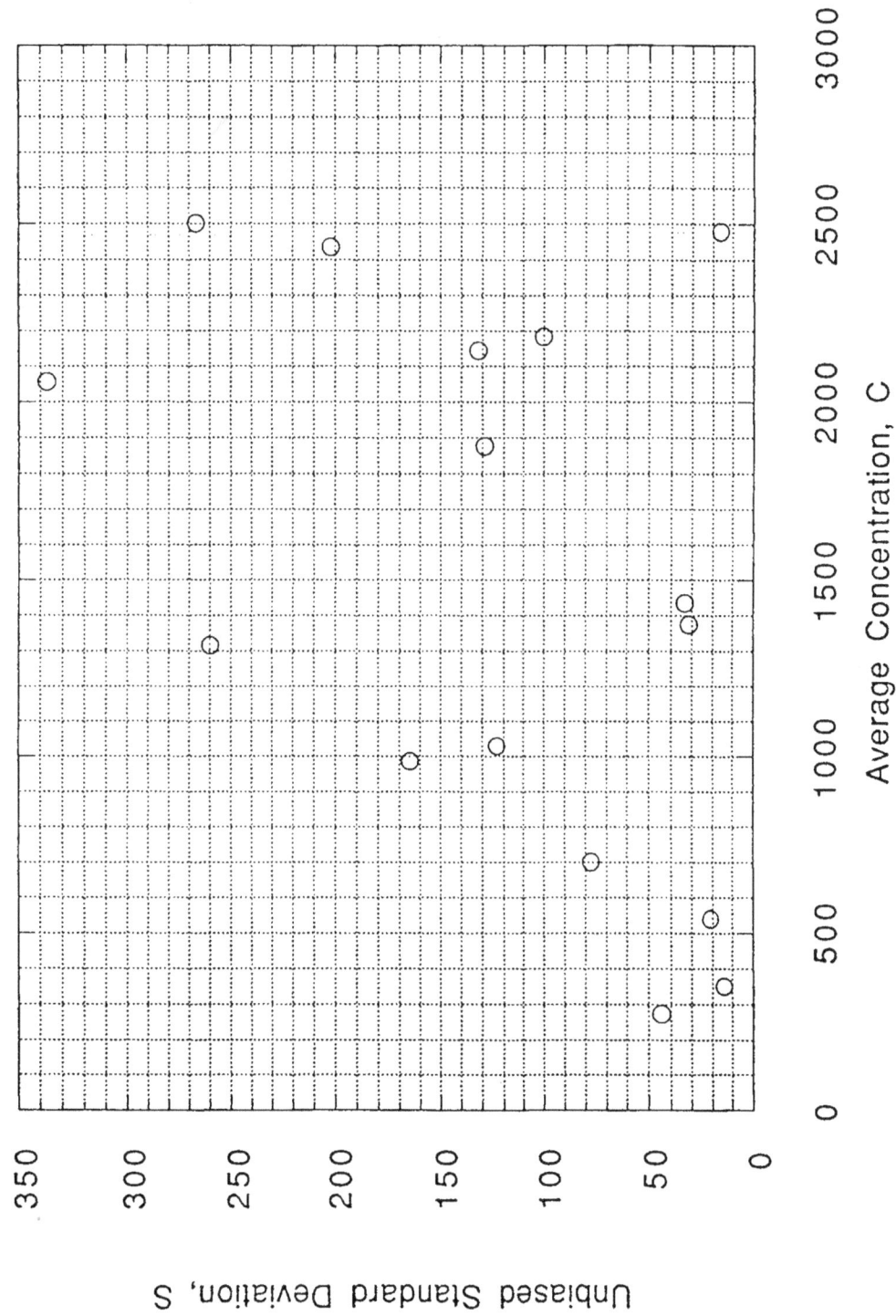

Figure 1
Simulated Data

Figure 2
Simulated Data

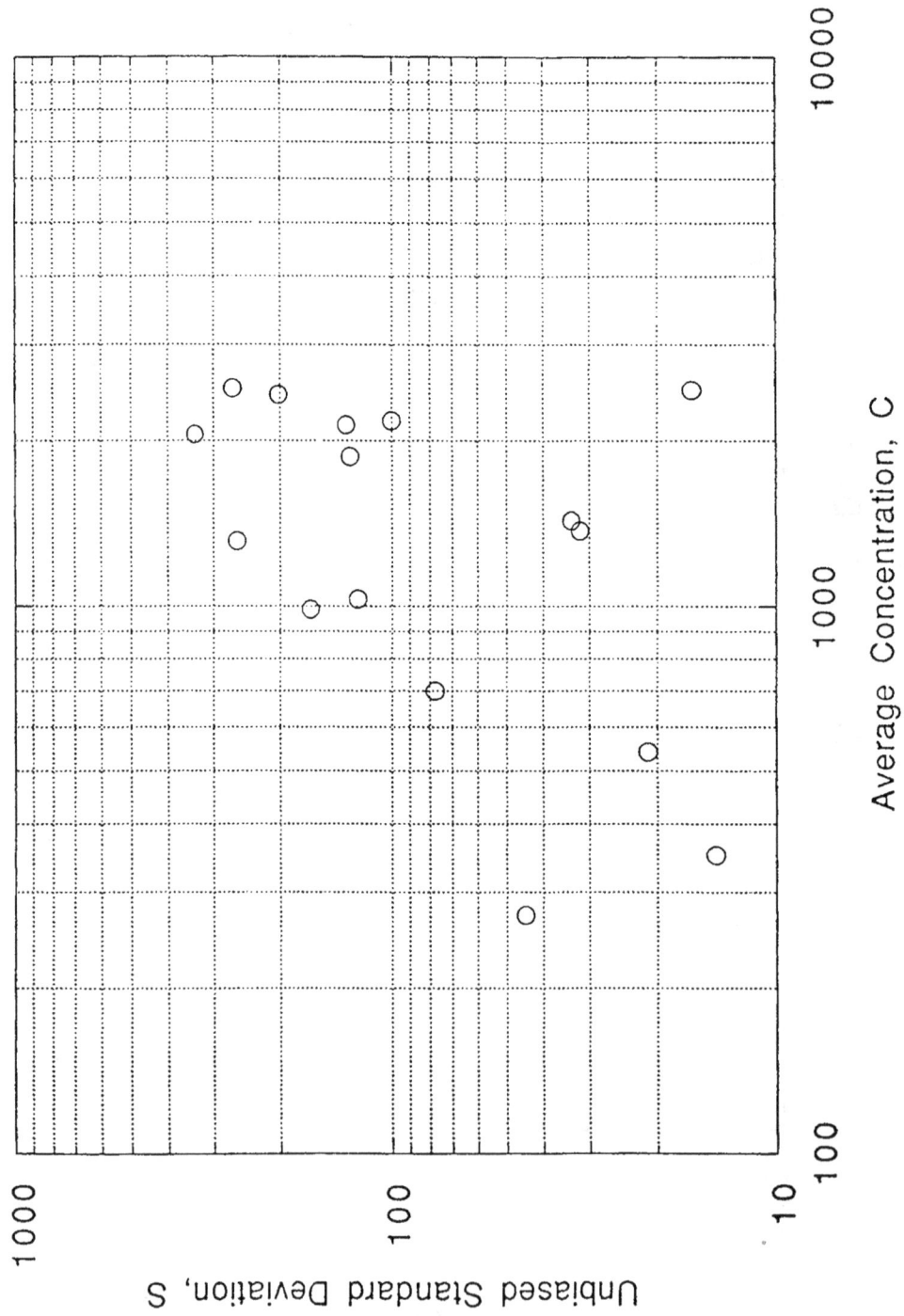

Average Concentration, C

Unbiased Standard Deviation, S

Table 3

```
                           T-CRIT 0.05 & 0.01 =  2.17 &  2.98
VARIABLES                  COEFFICIENTS    SE OF COEFF    T-RATIO
 0   Intercept                -0.80932
 1   Power Coefficient  p =   0.73084        0.34030        2.15
RESSUMSQ    STDDEV OF RES  DF     R-SQ
    12.16245      0.93207   14   0.2478
```

	In Ln Units				In Original Metrics		
	OBSVD	PRED	RESID	STD RES	Observed	Predicted	Diff.
1	5.107	4.230	0.878	0.94	165.25	68.69	96.56
2	4.610	4.810	-0.200	-0.21	100.48	122.73	-22.25
3	4.886	4.797	0.089	0.10	132.37	121.14	11.23
4	3.053	3.791	-0.738	-0.79	21.18	44.29	-23.11
5	5.586	4.909	0.677	0.73	266.78	135.55	131.23
6	2.664	3.474	-0.809	-0.87	14.36	32.25	-17.89
7	5.561	4.440	1.121	1.20	260.14	84.77	175.37
8	5.310	4.890	0.420	0.45	202.27	132.90	69.37
9	2.802	4.903	-2.101	-2.25**	16.48	134.67	-118.19
10	3.517	4.503	-0.986	-1.06	33.67	90.29	-56.62
11	4.353	3.978	0.375	0.40	77.70	53.42	24.28
12	3.463	4.472	-1.009	-1.08	31.90	87.52	-55.62
13	4.816	4.261	0.556	0.60	123.51	70.87	52.64
14	4.862	4.699	0.163	0.17	129.27	109.84	19.43
15	3.799	3.289	0.510	0.55	44.66	26.81	17.85
16	5.821	4.766	1.055	1.13	337.31	117.47	219.84

```
RES SUMSQ FROM REGRESSION =    12.16245
RES SUMSQ DIRECT =             12.16245
Average Observed =  122.3331
Average Predicted =  89.5749
Bias Correction Factor = 122.3331/89.5749 = 1.3657
```

$$Ln(S) = -0.8093 + 0.731 * Ln(C)$$

$$S = Exp(-0.8093) * C^{0.731}$$

$$S = 0.445 * C^{0.731}$$

In Table 3 the observed and predicted values of S and their residuals (differences between observed and predicted) are shown in Ln units, then in their original metrics. Sample calculation:

From Table 2, first row: C = 986.95

$$S_{pred} = Exp(-0.8093)^*C^{0.731} = 0.445^*C^{0.731} = 68.69 \quad \text{(Table 3)}$$

The average of the observed values is 122.3331; the average of the predicted values is 89.5749. The ratio of the average observed to average predicted is 122.3331/89.5749 = 1.3657. This offset or bias was caused by linearizing and fitting in terms of Ln(S) followed by conversion back to the original units.

The unbiased model for predicting σ from C is:

$$\text{Est } \sigma = 1.3657^*0.445^*C^{0.731} \quad \text{or} \quad \text{Est } \sigma = 0.608^*C^{0.731}$$

where Est σ is an unbiased estimate of sigma.

C	Est σ	RSD
1	0.608	60.8%*
10	3.27	32.7%*
100	17.62	17.6%
1000	94.82	9.5%

* extrapolation

By fitting all of the pollutant data to consistent models we can compare model coefficients for one pollutant against model coefficients for other pollutants.

How Good is This Model?

The answer to this question depends in part on what we intend to do with the model. Let's examine some of the statistics that come from the model-building process.

```
                              T-CRIT 0.05 & 0.01 = 2.17 & 2.98
VARIABLES                 COEFFICIENTS    SE OF COEFF    T-RATIO
 0  Intercept               -0.80932
 1  Power Coefficient   p = 0.73084        0.34030        2.15
RESSUMSQ    STDDEV OF RES   DF     R-SQ
   12.16245        0.93207  14   0.2478
```

The model coefficients are -0.8093 and 0.731. SE OF COEFF is a measure of how well we have estimated the coefficient, p. t = 0.73087/0.34031 = 2.15. A t-ratio larger about 2.0 implies we have detected a relationship between σ and C. A t-ratio > 2 (approx.) means that σ is not a constant, but is associated with C. However, t-ratios substantially larger than 2.0 are necessary to accurately estimate a model coefficient. The reference value of t at the 0.05 (or 95% probability level) is actually 2.15. See the line: T-CRIT 0.05 & 0.01 = 2.17 & 2.98. So our observed value of t (2.15 as compared to the reference value 2.17) is "right on the edge". The 95% confidence interval on the coefficient p is:

$$0.731 \pm 2.17*0.340 \quad \text{or from 0.00 to 1.47}$$

which suggests that the true coefficient could be as small as "zero" (implying no association between C and σ) or as large as 1.47. It would be misleading to report p = 0.731 without disclosing the uncertainty in this estimate. If we compare a power coefficient p for one pollutant to that of another, we must recognize that when the t-ratio for p is modest, the confidence interval on that value of p may be quite wide.

R-SQ (R-squared) = 0.2478 means this model explains 24.78% of the variation in the data. In this context "the data" is in terms of Ln(S). STDDEV OF RES is the standard deviations of residuals. This is the standard deviation of the differences between observed and predicted values in Ln(S) units. *This is not an estimate of σ.* STDDEV OF RES is a measure of *variation among values of Ln(S).* It is also an estimate of the standard deviation that we would find if we could run replicate tests under a fixed set of conditions *and report the variation among values of Ln(S).*

A logical extension to this would be to inquire "how good is a prediction made from this model?" This suggests we will calculate the unbiased estimate of σ (Est σ) for a fixed value of C.... then calculate a confidence interval on that predicted value. Before doing this, here are two values that we will need. One of these is the average of Ln(C) and the other is average of the Ln(S). Avg(LnC) = 7.1116. Avg(LnS) = 4.3881.

The confidence interval on Est σ is best found by re-stating the model in a different format:

Original Format: $Ln(Est\ \sigma) = -0.8093 + 0.731 \cdot Ln(C)$

New Format: $Ln(Est\ \sigma) = Avg(LnS) + 0.731 \cdot [Ln(C) - Avg(LnC)]$
New Format: $Ln(Est\ \sigma) = 4.3881 + 0.731 \cdot [Ln(C) - 7.1116]$

The variance of a prediction of Ln(σ) is:

$Var(Ln\sigma) = [StdDev\ of\ Res]^2/N + [SE(Coeff)]^2 \cdot [Ln(C) - 7.1116]^2$ Eq. 1

Table value of t
|
T-CRIT 0.05 & 0.01 = 2.17 & 2.98

VARIABLES	COEFFICIENTS	SE OF COEFF	T-RATIO
0 Intercept	-0.80932		
1 Power Coefficient	p = 0.73084	0.34030	2.15

RESSUMSQ	STDDEV OF RES	DF	R-SQ
12.16245	0.93207	14	0.2478

N = 16

t-ratio calculated from data

$Var(Ln\sigma) = [0.9321]^2/16 + [0.3403]^2 \cdot [Ln(C) - 7.1116]^2$

$Var(Ln\sigma) = 0.0543 + 0.1158 \cdot [Ln(C) - 7.1116]^2$

$SE(Ln\sigma) = \sqrt{Var(Ln\sigma)}$ This is the standard deviation of the predicted value of Lnσ, also called the standard error of prediction. See note at the bottom of page 14.

Here is the sequence for calculating a 95% confidence interval on σ:

 Step 1: Calculate the predicted value of Ln(Est σ) from either of:

 Ln(Est σ) = -0.8093 + 0.731*Ln(C)   or

 Ln(Est σ) = 4.3881 + 0.731*[Ln(C) - 7.1116]

 Step 2: Calculate the variance of Ln(Est σ) and the standard error of Ln(Est σ):

 $Var(Est\ \sigma) = 0.0543 + 0.1158*[Ln(C) - 7.1116]^2$

 SE(Est σ) = √Var(Est σ)   See footnote.

 Step 3: The 95% confidence interval on Ln(Est σ) is then

 Ln(Est σ) ± t*SE(Est σ) where t is the table or reference value of t at the 0.05 level of significance. t = 2.17 in this instance.

 Step 4: Revert to the original units and the unbiased estimates:

 Take antiLn of Est σ
 Take antiLn of the lower bound
 Take antiLn of the upper bound

 Multiply Est σ, the lower bound, and the upper bound by the bias correction factor 1.3657

Est σ and the 95% confidence limits on σ are shown in Table 4.

The expression *Standard Error* (Symbol: SE) means *the standard deviation of a statistic*. Literally, the standard deviation of a value calculated from data. This is standard notation; it is designed to keep issues about "the standard deviation" (calculated from "the data") separated from issues about the standard deviation of other quantities calculated from data.

<u>Table 4</u>

Average C	Lower Limit	Est σ	Upper Limit
10	0.090	3.269	118.079
20	0.249	5.425	118.174
50	0.946	10.600	118.756
100	2.583	17.594	119.850
200	6.979	29.203	122.137
500	24.797	57.058	131.288
1000	55.878	94.704	160.509
2000	84.433	157.189	292.637
5000	96.804	307.125	974.399

A graphical analysis of this table is presented in Fig. 3.

Minimizing the Width of the Confidence Interval

Equation 1 (repeated here) determines the width of the confidence interval on the estimated value of sigma, Est σ.

$$Var(Est\ \sigma) = [StdDev\ of\ Res]^2/N + [SE(Coeff)]^2*[Ln(C) - AvgLn(C)]^2 \quad Eq.\ 1$$

Whereas Est σ is a measure of variation due to sampling and analyzing, the StdDev of Res (standard deviation of residuals) *is a measure of the variation among individual values of Ln(S)*. Reducing the variation in sampling and analysis methods would reduce the magnitude of σ and would probably reduce StdDev of Res as well. A reduction in StdDev of Res would reduce the width of the confidence intervals.

N is the "number of tests" (number of duplicates, triplicates, quads, etc.); literally the number of rows in the data table from which we fit S vs C. Increasing N will decrease $[StdDev\ of\ Res]^2/N$ and will therefore decrease the width of the confidence interval.

SE(Coeff) is strongly influenced by the span of the data along the scale of C, the concentration of a pollutant. Varying C over a wide range will decrease SE(Coeff) and will therefore decrease the width of the confidence interval. Furthermore, if about one-half of the data are concentrated at a "low value of C" and one-half of the data are concentrated at a "high value of C", then SE(Coeff) will be minimized. SE(Coeff) is also reduced by increasing N and by reducing StdDev of Res.

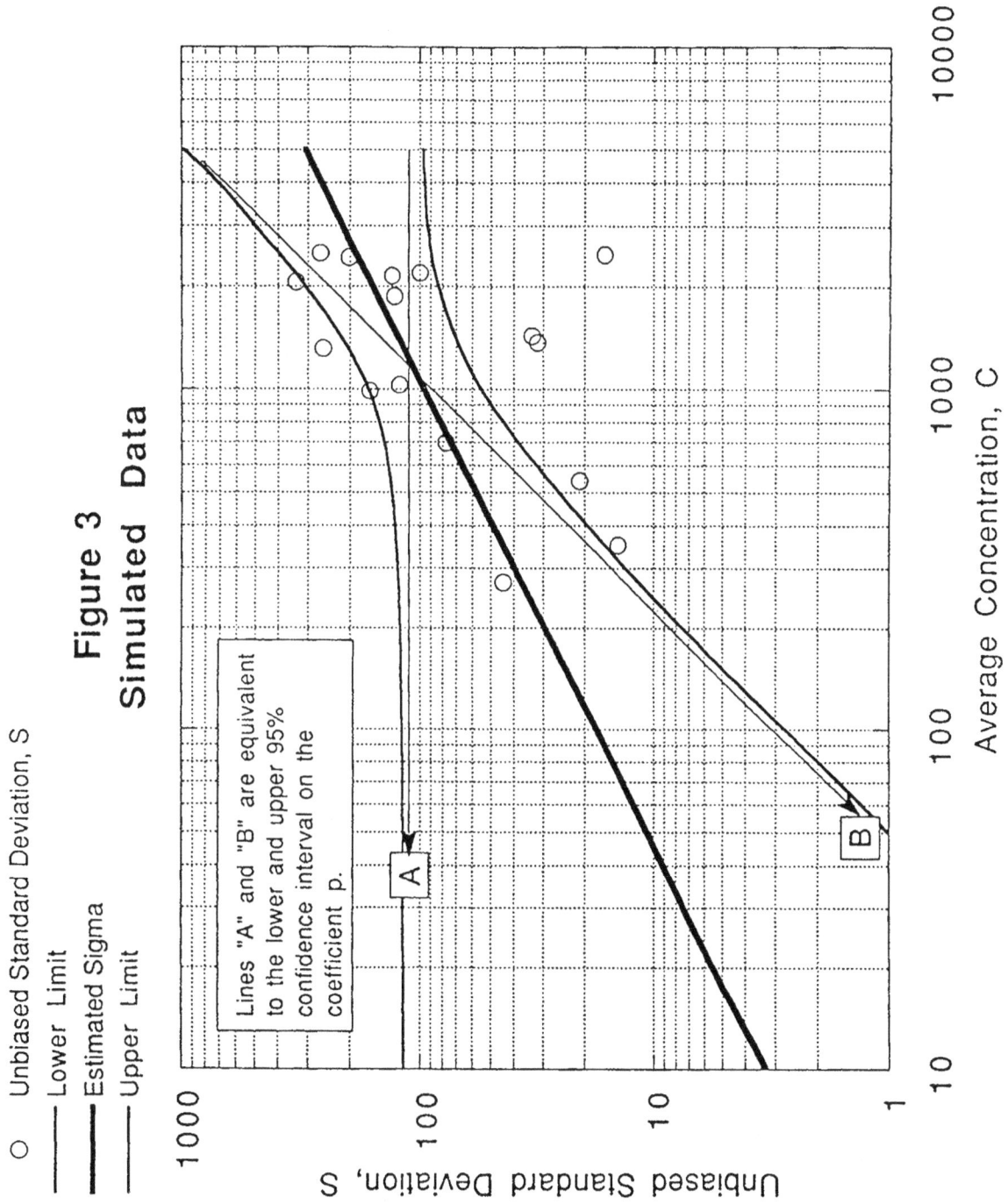

Figure 3
Simulated Data

○ Unbiased Standard Deviation, S
— Lower Limit
— Estimated Sigma
— Upper Limit

Lines "A" and "B" are equivalent to the lower and upper 95% confidence interval on the coefficient p.

Unbiased Standard Deviation, S

Average Concentration, C

$[Ln(C) - AvgLn(C)]^2$ is a function of where the prediction is being made along the scale of C. When a prediction is made at the centroid of $Ln(C)$ then $Ln(C) = AvgLn(C)$ and this term goes to zero. When this is so, then $Var(Est \sigma) = [StdDev of Res]^2/N$. This means that the width of the confidence interval will be minimized when predictions are made at the centroid of $Ln(C)$.

The best way to visualize this is with an example in which data are concentrated at two points on the scale of C. The data in Table 5 are simulated data from the same source as the foregoing example.

Table 5

Sample	Data		Avg, C	S	Unbiased S
1	243.3	253.6	248.45	7.28	9.12
2	237.3	233.7	235.50	2.55	3.20
3	151.1	143.5	147.30	5.37	6.73
4	250.4	216.6	233.50	23.90	29.95
5	173.1	207.6	190.35	24.40	30.57
6	141.8	159.0	150.40	12.16	15.24
7	233.2	208.4	220.80	17.54	21.98
8	139.9	153.6	146.75	9.69	12.14
9	1733.4	1828.4	1780.90	67.18	84.18
10	2022.6	1754.9	1888.75	189.29	237.18
11	2127.9	2014.9	2071.40	79.90	100.11
12	2064.0	2047.6	2055.80	11.60	14.53
13	2016.6	1934.1	1975.35	58.34	73.10
14	1877.2	1767.2	1822.20	77.78	97.46
15	2161.3	2492.2	2326.75	233.98	293.18
16	2003.9	1808.6	1906.25	138.10	173.04

The relationship between S and C is shown in Figures 4 and 5. Table 6 is the least squares analysis of the data in Table 5.

Note Lines "A" and "B" in Fig. 3. These are equivalent to the lower and upper 95% confidence interval on the slope in Ln-Ln coordinates; i. e., on the model coefficient p. Recall that the lower limit on p was 0.00; hence "A" is horizontal. This may aid in seeing how the imprecision in estimating the slope impacts the confidence interval on predicted values of sigma. This imprecision in estimating the slope is reported as SE of Coeff.

Figure 4
Simulated Data

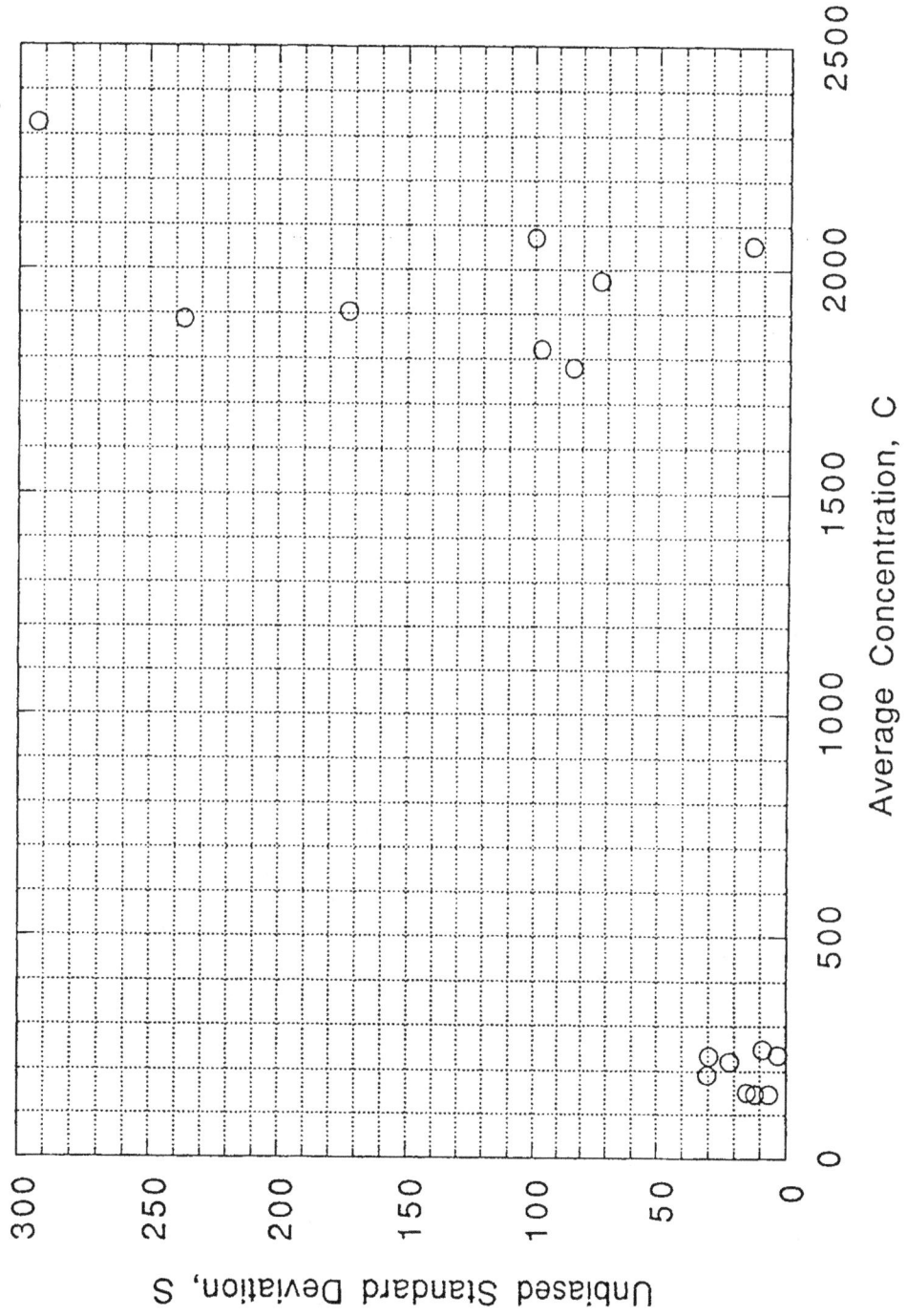

Average Concentration, C

Unbiased Standard Deviation, S

Figure 5
Simulated Data

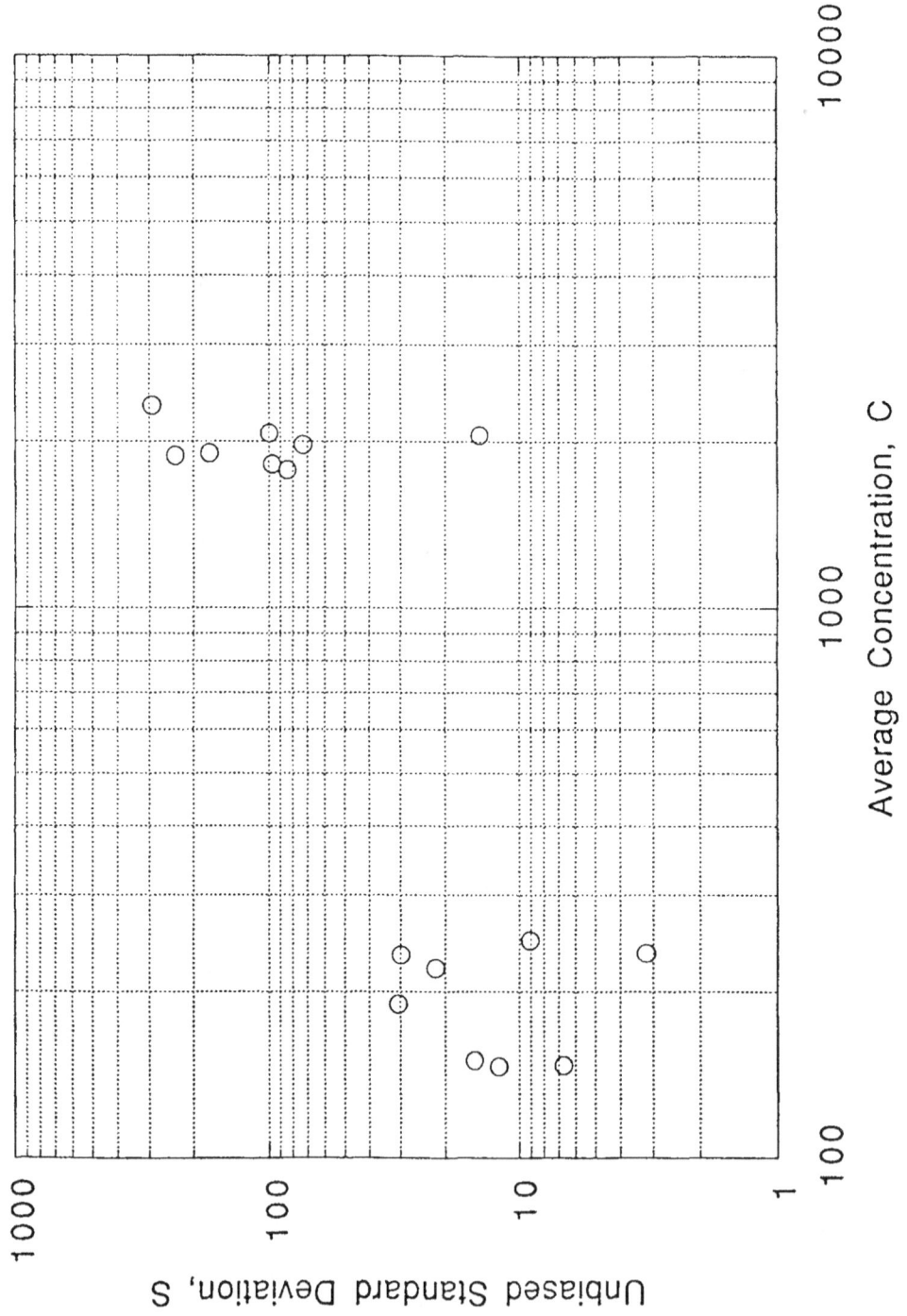

Average Concentration, C

Unbiased Standard Deviation, S

Table 6

Table value of t
|
T-CRIT 0.05 & 0.01 = 2.17 & 2.98

VARIABLES	COEFFICIENTS	SE OF COEFF	T-RATIO
0 Intercept	-2.01370		
1 Power Coefficient p ≈ 0.87111		0.18436	4.73

RESSUMSQ	STDDEV OF RES	DF	R-SQ	
10.51455	0.86663	14	0.6146	

|
StdDev of Res   N = 16

t-ratio calculated
from data

	In Ln Units				In Original Metrics		
	OBSVD	PRED	RESID	STD RES	Observed	Predicted	Diff.
1	2.210	2.791	-0.580	-0.67	9.12	16.29	-7.17
2	1.163	2.744	-1.581	-1.82	3.20	15.55	-12.35
3	1.907	2.335	-0.429	-0.49	6.73	10.33	-3.60
4	3.400	2.737	0.663	0.76	29.95	15.43	14.52
5	3.420	2.559	0.861	0.99	30.57	12.92	17.65
6	2.724	2.353	0.371	0.43	15.24	10.52	4.72
7	3.090	2.688	0.402	0.46	21.98	14.70	7.28
8	2.497	2.332	0.164	0.19	12.14	10.30	1.84
9	4.433	4.506	-0.073	-0.08	84.18	90.60	-6.42
10	5.469	4.558	0.911	1.05	237.18	95.36	141.82
11	4.606	4.638	-0.032	-0.04	100.11	103.34	-3.23
12	2.676	4.631	-1.955	-2.26**	14.53	102.66	-88.13
13	4.292	4.597	-0.305	-0.35	73.10	99.16	-26.06
14	4.579	4.526	0.053	0.06	97.46	92.42	5.04
15	5.681	4.739	0.941	1.09	293.18	114.36	178.82
16	5.154	4.566	0.588	0.68	173.04	96.13	76.91

RES SUMSQ FROM REGRESSION = 10.51455
RES SUMSQ DIRECT =   10.51456
Average Observed =   75.1069
Average Predicted = 56.2545
Bias Correction Factor = 75.1069/56.2545 = 1.3351

$$\text{Est } \sigma = 1.3351 * \text{Exp}(-2.0137) * C^{0.871}$$

$$\text{Est } \sigma = 1.3351 * 0.1335 * C^{0.871}$$

$$\text{Est } \sigma = 0.178 * C^{0.871}$$

The 95% confidence interval on p is:

$$0.871 \pm 2.17*0.184 \quad \text{or from} \quad 0.47 \text{ to } 1.27$$

an improvement over 0.00 to 1.47 in the previous example.

Averages of S and C in Ln units: $S_{avg} = 3.58126$ $C_{avg} = 6.42282$

Original Format: $Ln(Est \ \sigma) = -2.0137 + 0.8712*Ln(C)$

New Format: $Ln(Est \ \sigma) = S_{avg} + 0.8712*[Ln(C) - C_{avg}]$
New Format: $Ln(Est \ \sigma) = 3.58126 + 0.8712*[Ln(C) - 6.42282]$

The variance of a prediction is (Eq. 1 is repeated here):

$$Var(Est \ \sigma) = [StdDev \ of \ Res]^2/N + [SE(Coeff)]^2*[Ln(C) - Avg \ of \ Ln(C)]^2 \quad Eq. \ 1$$

$$Var(Est \ \sigma) = [0.86663]^2/16 + [0.18436]^2*[Ln(C) - 6.42282]^2$$

$$Var(Est \ \sigma) = 0.04694 + 0.0344*[Ln(C) - 6.42282]^2$$

Note that $[SE(Coeff)]^2 = 0.0344$ compared to 0.1158 in the previous example. By concentrating data at the extremes of the experimental space the width of the confidence interval has been reduced significantly. Est σ and the 95% confidence limits on σ are shown in Table 7.

Table 7

Average C	Lower Limit	Est σ	Upper Limit
10	0.236	1.324	7.421
20	0.564	2.422	10.400
50	1.765	5.381	16.402
100	4.125	9.843	23.484
200	9.375	18.004	34.577
500	24.812	40.000	64.485
1000	43.978	73.167	121.723
2000	68.641	133.836	260.951
5000	113.262	297.342	780.600

A graphical analysis of Table 7 is presented in Figure 6

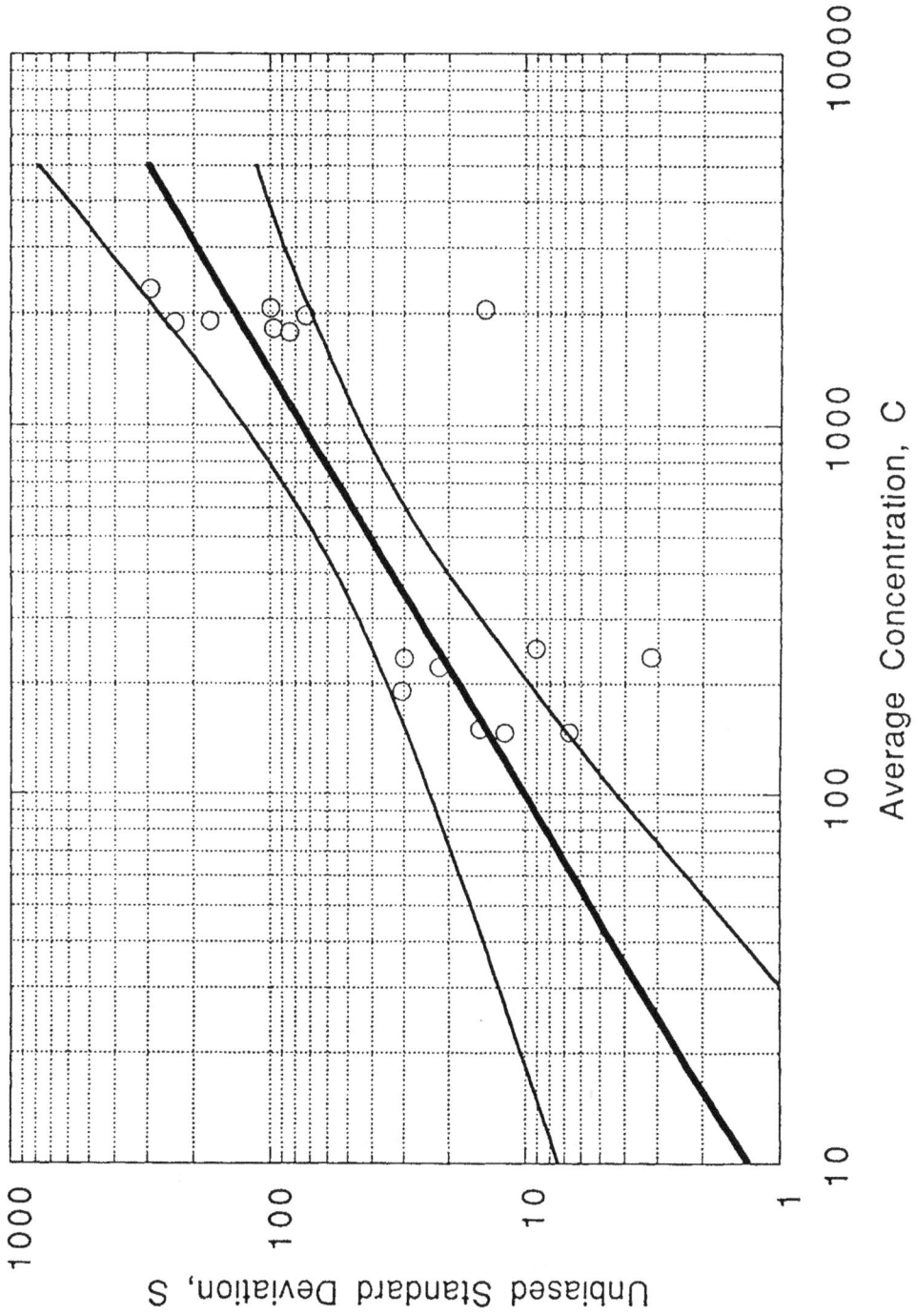

Figure 6
Simulated Data

The simulated data used in these examples (Tables 2 and 5) came from this model: $\sigma = 0.28 \cdot C^{0.8}$. The value of C was varied randomly in Table 2 and varied randomly around two points on the scale of C in Table 5. True values of σ were calculated from C; from a source with mean C and standard deviation σ, two random normal numbers were generated; those numbers are the data. This process was repeated 16 times. When C = 100 the true RSD = 11.1%; when C = 1000 the true RSD = 7.0%. These levels of RSD are typical of those encountered in the analysis of the actual pollutant data.

The model derived from data in the first simulation (Table 2) is Est $\sigma = 0.608 \cdot C^{0.731}$ The 95% confidence interval on p is 0.00 to 1.47.

The model derived from data in the second simulation (Table 5) is Est $\sigma = 0.178 \cdot C^{0.871}$ The 95% confidence interval on p is 0.47 to 1.27.

With this in mind, here are some important conclusions.

- Even when simultaneously sampled data come from a "perfect" situation such that:

  <> the underlying model is exact and the data are contaminated only by random variation

  <> there are no concerns about sample contamination, selective or biased sampling of particles, or other "special causes"

  finding a relationship between Est σ and the average concentration C yields model coefficients and predictions that are subject to statistical uncertainty.

- The evidence of this uncertainly is illustrated by the fact that two sets of data from a "perfect situation" produce models that have different coefficients and therefore different estimates of Est σ as a function of the average concentration C.

- SE(Coeff) is a major contributor to wide confidence intervals on Est σ. Although not discussed in detail here, SE(Coeff) is strongly influenced by (1) the allocation of experimental· points along the scale of C; (2) the amount of data; and (3) the inherent variation in the data.

Weighted Regression

Although the majority of the ReMap data were simultaneous paired samples with N = 2, there were instances of triads, quads, and even some octets. Recall that calculated values of S are being treated as "data". When those values are derived from varying amounts of information, they should be weighted in accordance with the amount of information in each value of S.

As a rule, the quantity to be minimized with weighted least squares is $Q = $ Sum of $W_i^*(S_i - \text{pred } S_i)^2$ where Si is an observed standard deviation; pred Si is the predicted value of that standard deviation; and W_i is the weight assigned to the i-th observation. The weights should be inversely proportional to the variance of the residuals. As an approximation, Minimize $Q = $ Sum of $(1/\text{Var } S_i)^*(S_i - \text{pred } S_i)^2$

The variance of S_i is (an approximation) proportional to $1/[2(N - 1)]$. Thus the relative weights to be assigned to each value of S is $2(N - 1)$ where N is the number of observations used to calculate a given value of S. This is the same as 2(df). The weight assigned to each observed value of S can be reduced to just the number of degrees of freedom associated with that value of S. If N = 2 the weight is 1. If N = 3 the weight is 2, etc.

The calculations for weighted regression are similar to those for unweighted regression (pages 7 - 23). But the calculations for weighted regression are substantially more complex. In particular, calculating the confidence intervals requires knowledge of matrix operations. For these reasons we recommend that software designed for weighted regression be used for this purpose.

Errors, Outliers, and Mavericks

Questionable data is given a diversity of names, such as mavericks, fliers, outliers, sports, and blunders. These aberrations can be caused by contamination of samples, switched or mislabeled samples, faulty equipment or reagents, key entry errors, and a host of other events. Such data is a source of frustration, and often causes pointless discussions and wasted effort. Some statistical criteria are available to assist in learning whether or not the largest or smallest observation is significantly far removed from the main body of the data.

In this section we will address this matter from two perspectives. The first of these is an omnibus examination of all of the data in one pass. The other is more focused, and addresses only one triplicate or quad.

Table 8 was derived from Table 5. The data were broken into two categories... low C and high C... with the expectation that variation will be constant (or virtually constant) at low C and that variation will be constant at high C. Of course we expect that variation will change between low C and high C, which was the point to breaking the table into these two categories.

Table 8

Sample	Data		Avg, C	Range
1	243.3	253.6	248.45	10.3
2	237.3	233.7	235.50	3.6
3	151.1	143.5	147.30	7.6
4	250.4	216.6	233.50	33.8
5	173.1	207.6	190.35	34.5
6	141.8	159.0	150.40	17.2
7	233.2	208.4	220.80	24.8
8	139.9	153.6	146.75	13.7

Avg Range = 18.19

Sample	Data		Avg, C	Range
9	1733.4	1828.4	1780.90	95.0
10	2022.6	1754.9	1888.75	267.7
11	2127.9	2014.9	2071.40	113.0
12	2064.0	2047.6	2055.80	16.4
13	2016.6	1934.1	1975.35	82.5
14	1877.2	1767.2	1822.20	110.0
15	2161.3	2492.2	2326.75	330.9
16	2003.9	1808.6	1906.25	195.3

Avg Range = 151.35

The range is shown for each simultaneous sample. The average range is reported for each of the two groups.

In Statistical Process Control (SPC) methods it is customary to test whether any of these ranges are abnormal. With duplicates, triads, and quads we only test for abnormally large ranges; testing for abnormally small ranges is not meaningful. The factors in Table 9 are used to determine whether any of the ranges are abnormally *large*.

<u>Table 9</u>

Sample Size, n	D_4
2	3.267
3	2.575
4	2.282
5	2.115

Step 1: The data used to calculate ranges were duplicates. The sample size is n = 2. D_4 = 3.267

Step 2: Multiply the average range by D_4 for each category.
3.267 x 18.19 = 59.4   3.267 x 151.35 = 494.5

Step 3: Compare the individual ranges against these limits. When C is low, ranges larger than 59.4 are suspect. When C is high, ranges larger than 494.4 are questionable.

All of the ranges passed the test. There is no evidence for excluding any of the data.

The Avg Ranges were calculated from only eight ranges. In practice we should average 10 or more ranges within a category.

Now let's consider how this can be extended to a case in which the data was not concentrated at two points on the scale of C. Table 10 is derived from Table 2. Table 10 is ranked on C, and then broken into three categories.

Table 10

Sample	Data		Avg, C	Range
15	247.2	297.6	272.40	50.4
6	358.9	342.7	350.80	16.2
4	529.4	553.3	541.35	23.9
11	743.6	655.9	699.75	87.7
1	893.7	1080.2	986.95	186.5

72.94 = Avg Range

13	960.3	1099.7	1030.00	139.4
7	1463.0	1169.4	1316.20	293.6
12	1356.8	1392.8	1374.80	36.0
10	1415.8	1453.8	1434.80	38.0
14	1949.0	1803.1	1876.05	145.9

130.58 = Avg Range

16	1866.4	2247.1	2056.75	380.7
3	2070.3	2219.7	2145.00	149.4
2	2240.4	2127.0	2183.70	113.4
8	2549.1	2320.8	2434.95	228.3
9	2488.9	2470.3	2479.60	18.6
5	2351.1	2652.2	2501.65	301.1

238.22 = Avg Range

The rules are applied as before.

Step 1: The data used to calculate ranges were duplicates.
The sample size is n = 2. $D_4 = 3.267$

Step 2: Multiply the average range by D_4 for each category.
3.267 x 72.94 = 238.3
3.267 x 130.58 = 426.6
3.267 x 238.22 = 778.3

Step 3: Compare the individual ranges against these limits
category-by-category as before.

All of the ranges passed the test. There is no evidence for excluding any of the data.

This is presented as an example. In practice we should average 10 or more ranges when using this procedure.

If there are gaps in the ranked values of C, then break the data into categories at those points. There is no requirement that the same amount of data will be in each category.

The extension of this to triplicates and quads only requires using other values of D_4.

If we examine only one set or triplicates or quads, then Dixon's-r procedure is appropriate.[*]

Consider the following simultaneously sampled data:

22.3   29.4   49.1   28.2

Step 1: Rank the data from smallest to largest:
23.8   28.2   29.4   49.1

Step 2: Calculate r = (49.1 - 29.4)/(49.1 - 23.8)
r = 0.779

Table 11

n	$P_{0.95}$	$P_{0.99}$
3	0.941	0.988
4	0.765	0.889
5	0.642	0.780

Step 3: The sample size n = 4.

Step 4: The calculated value r = 0.779 is larger than $P_{0.95}$ ($P_{0.95}$ = 0.765). The evidence suggests that 49.1 is not consistent with the remainder of the data.

If the calculated value of r exceeds the table value at $P_{0.99}$ then the evidence is even stronger.

[*] Dixon, W. J. and Massey, F. J.; *Introduction to Statistical Analysis*, 3rd Ed.; McGraw-Hill; 1969. pages 328 - 330.

Dixon's-r is of the form: r = (Distance between the largest and its nearest neighbor)/(Full Range of the data.) This can be arranged to inquire about the status of a number that is unusually low when compared to the remainder of the data.

Example: 121 179 185 193

$$r = (179 - 121)/(193 - 121) = 0.805$$

This is larger than the table value of r (0.765) with n = 4.
So the 121 is inconsistent with the remainder of the data.

Dixon's-r cannot be used with n = 2 data. It assumes the data came from a normal distribution. This is a reasonable assumption with simultaneously sampled concentration data.

See Table 2 test 9 and Fig. 2. S = 16/48 seems to be unusually low. This is also appears as a large residual in Table 3. But recognize that with small samples (especially duplicates and triads) is is possible for the standard deviation to be very low. In the case of duplicates S may even be "zero" occasionally. So even though these may seem unusual, they are not.

Fitting Standard Deviation vs. Concentration Data to Alternative Models

A power-law model was used throughout this report because the data will not support more than two model coefficients.

There may be instances in which we will need to entertain other model forms. For example, please draw a line or curve to express the relationship between S and C in Fig. 7. Do this before proceeding.

The relationship between S and C does not seem to be linear in Ln-Ln coordinates. It is reasonable to consider an extension to the power-law model, perhaps $S = a + kC^p$. In this alternative model S will approach a when C = 0. This suggests there is a lower bound on σ.

Before we begin setting up the tools to estimate a, k, and p in this extended model, we should work through the following process.

1. Test the data to determine whether the appearance of curvature (in Ln-Ln coordinates) is "real" or simply due to chance variation in the data. If the perceived curvature can be attributed to random variation in the data... and not a systematic offset from a straight line relationship... then attempting to fit these data to a model more complex than $S = kC^p$ will be misleading and disappointing. Apply the test for curvature after transforming to Ln-Ln coordinates, of course.

2. The most direct way to test for curvature is to fit the data to the usual power-law model... $Ln(S) = Ln(k) + pLn(C)$... then ask whether there is evidence of "lack-of-fit". This is easily done by testing to see whether the data will support adding $b[Ln(C)]^2$ to the model. If the t-ratio associated with b is larger than the appropriate reference value of the t-ratio (as a rule, larger than 2), then there is evidence of curvature; the simple power-law model is not adequate. If the t-ratio associated with b is notably smaller than 2, there is no firm evidence of curvature; the power-law model is adequate; attempting to fit the data to a more complex model to account for curvature is pointless and misleading.

3. The purpose of (2) above is not to build a completed model. The purpose is to test for the presence of curvature beyond that which is accommodated by the power-law model. If curvature is detected (t-ratio for b is larger than, say, 2), then we may be justified in considering an alternative model. $S = a + kC^p$ would be a candidate.

Figure 7

Is The Line Curved or Straight?

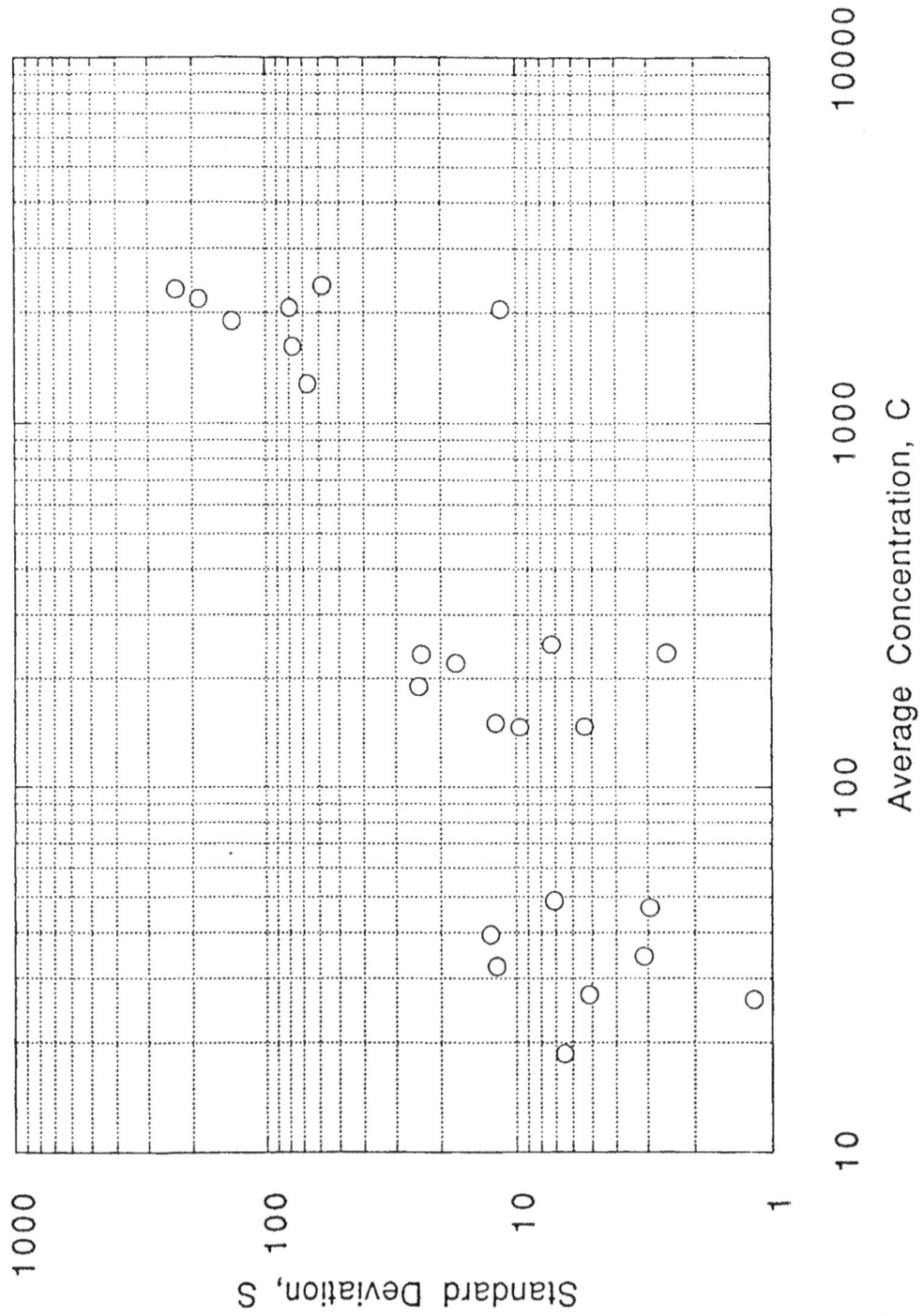

This process was followed to make that judgement about these data.

C	S	C	S
248.5	7.28	2375.0	58.34
235.5	2.55	1622.0	77.78
147.3	5.37	2326.0	234.0
233.5	23.9	1906.0	138.1
190.4	24.4	18.6	6.48
150.4	12.16	26.3	1.19
220.8	17.54	27.0	5.18
146.7	9.69	32.4	12.06
1281.0	67.18	34.6	3.12
2189.0	189.29	39.6	12.82
2071.0	79.9	46.7	2.96
2056.0	11.6	48.9	7.10

In Fig. 8 the power-law model explained 66.5% of the variation in Ln(S). In Fig. 9 the model was extended by adding a quadratic coefficient as suggested on page 30. This extended model explains 69.2% of the variation in Ln(S). R-sq will always increase when we add another model coefficient; this increase in R-sq does not prove that the quadratic term is important. The t-ratio for the quadratic term is an indication of whether that term is a worthy addition to the model.

The t-ratio for the quadratic coefficient is only 1.36; far below the reference value of 2.08 (with 21 df, 0.05 level). This low t-ratio means we do not have sufficient evidence to support adding "curvature" to the power-law model $S = kC^p$. What we perceived as "real" curvature could have come from a straight-line relationship between Ln(σ) and Ln(C).

This does not mean we have "proved" that a straight-line relationship prevails in Ln-Ln coordinates. It simply means that the evidence in favor of a more complex relationship is weak. There is little justification for adding a quadratic term to the original model; there is little justification for pursuing an alternative model such as $S = a + kC^p$.

It is possible that with additional data we may learn that curvature is actually present, and that an extended model form was justified. But attempting to build a model more complex than $S = kC^p$ is not appropriate with the existing data. If we pursue this... if we build an extended model (say, $S = a + kC^p$)... then the coefficients in that model will be poorly estimated; the confidence intervals on those coefficients will be extremely wide; and we have gained nothing more than the satisfaction of explaining a little more of the variation in the data.

Figure 8   page 33

STEP NO. 2   Determinant = 1.0000
VAR. 1 GOING IN   T-CRIT 0.05 & 0.01 = 2.07 & 2.80
VARIABLES   COEFFICIENTS   SE OF COEFF   T-RATIO
 0   Intercept   2.77468
 1   average conc   0.68217   0.10316   6.61
RESSUMSQ   STDDEV OF RES   DF   R-SQ
     16.02539   0.85348   22   0.6653 <--- 66.5%

$$Ln(S) = 2.775 + 0.682*(Ln(C) - 5.44)$$

		In Ln Units			In Original Units		
	OBSVD	PRED	RESID	STD RES	Observed	Predicted	Diff
1	1.985	2.826	-0.841	-0.99	7.28	16.88	-9.60
2	0.936	2.789	-1.853	-2.17	2.55	16.27	-13.72**
3	1.681	2.469	-0.789	-0.92	5.37	11.82	-6.45
4	3.174	2.784	0.390	0.46	23.90	16.18	7.72
5	3.195	2.644	0.550	0.64	24.40	14.08	10.32
6	2.498	2.484	0.015	0.02	12.16	11.98	0.18
7	2.864	2.746	0.119	0.14	17.54	15.57	1.97
8	2.271	2.467	-0.196	-0.23	9.69	11.78	-2.09
9	4.207	3.945	0.263	0.31	67.18	51.67	15.51
10	5.243	4.310	0.933	1.09	189.29	74.47	114.82
11	4.381	4.273	0.108	0.13	79.90	71.71	8.19
12	2.451	4.268	-1.817	-2.13	11.60	71.35	-59.75**
13	4.066	4.366	-0.300	-0.35	58.34	78.73	-20.39
14	4.354	4.106	0.248	0.29	77.78	60.70	17.08
15	5.455	4.352	1.104	1.29	234.00	77.62	156.38
16	4.928	4.216	0.712	0.83	138.10	67.76	70.34
17	1.869	1.058	0.811	0.95	6.48	2.88	3.60
18	0.174	1.294	-1.120	-1.31	1.19	3.65	-2.46
19	1.645	1.312	0.333	0.39	5.18	3.71	1.47
20	2.490	1.436	1.054	1.23	12.06	4.21	7.85
21	1.138	1.481	-0.343	-0.40	3.12	4.40	-1.28
22	2.551	1.573	0.978	1.15	12.82	4.82	8.00
23	1.085	1.686	-0.601	-0.70	2.96	5.40	-2.44
24	1.960	1.717	0.243	0.28	7.10	5.57	1.53

RES SUMSQ FROM REGRESSION = 16.02539 RES SUMSQ DIRECT = 16.02538
Average Observed = 42.0829 Average Predicted = 29.2997

Figure 9   page 34

STEP NO. 3   Determinant = 0.9685
VAR. 2 GOING IN   T-CRIT 0.05 & 0.01 = 2.08 & 2.82
VARIABLES   COEFFICIENTS   SE OF COEFF   T-RATIO
 0 Intercept   2.46120
 1   average conc   0.65730   0.10285   6.39
 2   quadratic   0.10992   0.08073   1.36 <--- Note
RESSUMSQ   STDDEV OF RES   DF   R-SQ
    14.72537   0.83738   21   0.6924   69.2% - 66.5% = 2.7%

$$Ln(S) = 2.461 + 0.657*(Ln(C) - 5.44) + 0.11*(Ln(C) - 5.44)\text{-sq}$$

| | \multicolumn{4}{c}{In Ln Units} | \multicolumn{3}{c}{In Original Units} |
	OBSVD	PRED	RESID	STD RES	Observed	Predicted	Diff
1	1.985	2.511	-0.526	-0.63	7.28	12.32	-5.04
2	0.936	2.476	-1.539	-1.84	2.55	11.89	-9.34
3	1.681	2.189	-0.508	-0.61	5.37	8.93	-3.56
4	3.174	2.470	0.704	0.84	23.90	11.82	12.08
5	3.195	2.340	0.855	1.02	24.40	10.38	14.02
6	2.498	2.201	0.297	0.36	12.16	9.03	3.13
7	2.864	2.433	0.431	0.51	17.54	11.40	6.14
8	2.271	2.187	0.084	0.10	9.69	8.91	0.78
9	4.207	3.912	0.295	0.35	67.18	50.01	17.17
10	5.243	4.498	0.745	0.89	189.29	89.83	99.46
11	4.381	4.434	-0.054	-0.06	79.90	84.31	-4.41
12	2.451	4.426	-1.975	-2.36	11.60	83.61	-72.01**
13	4.066	4.593	-0.526	-0.63	58.34	98.76	-40.42
14	4.354	4.162	0.191	0.23	77.78	64.23	13.55
15	5.455	4.568	0.887	1.06	234.00	96.38	137.62
16	4.928	4.341	0.587	0.70	138.10	76.75	61.35
17	1.869	1.503	0.366	0.44	6.48	4.50	1.98
18	0.174	1.552	-1.378	-1.65	1.19	4.72	-3.53
19	1.645	1.557	0.088	0.10	5.18	4.75	0.43
20	2.490	1.595	0.895	1.07	12.06	4.93	7.13
21	1.138	1.610	-0.472	-0.56	3.12	5.00	-1.88
22	2.551	1.645	0.906	1.08	12.82	5.18	7.64
23	1.085	1.692	-0.607	-0.72	2.96	5.43	-2.47
24	1.960	1.706	0.254	0.30	7.10	5.51	1.59

RES SUMSQ FROM REGRESSION = 14.72537   RES SUMSQ DIRECT = 14.72535
Average Observed = 42.0829   Average Predicted = 32.0234

The foregoing data can be fit directly to a model of the form $S = a + kC^p$ using the following technique. Set p to a constant (say, 0.6) and fit $S = a + kC^{0.6}$. Since this model is linear in the coefficients a and k, ordinary least squares methods can be used to estimate those coefficients. Repeat this process for p = 0.8, 1.0, etc. When this is done:

p	a	k	Residual Sum Sq
0.6	−10.27	1.24	40.52
0.8	−1.61	0.25	39.96
1.0	2.95	0.053	39.66
1.2	5.60	0.011	39.50
1.4	7.26	0.0024	39.43
1.6	8.39	0.00052	39.41

The coefficient a is an estimate of σ when C = 0. Negative values of a imply that σ becomes negative as the average concentration approaches zero. So p < 1 is certainly not acceptable.

The residual sum of squares... a direct measure of how well a model explains the variation in S, is insensitive to the least squares combinations of a, k, and p. The estimates of all three coefficients are highly correlated. There is no unique combination of coefficients that minimizes the residual sum of squares; one combination is as good as any other. This means that the confidence intervals on the model coefficients are very wide. This will happen when the data is not capable of accurately estimating the coefficients in the proposed model. In a favorable situation the residual sum of squares will have a "sharp" well-defined minimum.

The underlying chemistry or physics may suggest that σ converges to a limit greater than zero; Fig. 7 suggests this. What harm is done by fitting the S vs. C data to a model... viz., $S = a + kC^p$... that supports this theory about the lower bound on σ? That depends on the definition of harm. The prior analysis showed there is insufficient evidence to support "curvature" in Ln-Ln coordinates. If an alternative model is used to claim that the data follows a certain theory, when in fact the model coefficient that would support that theory is not supported by the data, then that is a poorly founded claim. The resolution to this is to get data that will properly test the theory; in this instance, get data at very low levels of C. *In any event, the confidence intervals on a, k, and p should be reported.*

The Performance of the ReMAP Process

The ReMAP process involves establishing relationships between estimates of sigma... sample standard deviations, S, each with very few degrees of freedom... and average values of concentration, those averages derived from small samples of data. Although the individual steps in the process can be found in textbooks, the total process is not a standard statistical procedure. It is therefore appropriate to test and verify that the ReMAP process performs as expected.

Monte Carlo simulations were run to measure the performance of the ReMAP process. *By performance we mean how frequently the confidence intervals on sigma encompass sigma*. These simulations were designed to verify that statements of the form "the probability is 95% that this interval has encompassed sigma" are accurate.

The simulations are comprised of the following steps.

1. Generate data that resembles actual data. The model used to generate the simulated data was Sigma = $0.2 * C^{0.8}$ This resembles models cited in the main body of this report. The simulated data were clustered near the ends of a concentration scale (C near 1000 and C near 4000); in some simulations they were scattered randomly along the scale of C. By clustered we mean the data were in a band approximately 500 to 1500 and in another band approximately 3000 to 5000 on the concentration scale.

2. Once a "true" value of C had been selected (randomly within a band) on the scale of concentration, the corresponding true value of sigma was calculated from Sigma = $0.2 * C^{0.8}$. Two simulated data were then generated from a random normal source with mean C and standard deviation sigma. The Box-Muller method was used to generate the random normal deviates

3. Steps 1 - 2 were repeated (say) 20 times to generate 20 sets of simultaneous samples, each comprised of a pair (N = 2) of data.

4. The sample average and sample standard deviation S were calculated for each pair of data. This value of S was multiplied by the small sample correction factor 1.253 to obtain unbiased estimates of sigma.

5. The (sample average, S)-combinations were used to build a model of the form Ln(S) = Ln(K) + p*Ln(C). This equation was used to predict Ln(S) for each of the 20 combinations. Estimates of Ln(S) were converted to estimates of S by Exp(Ln(S)).

6. As in the ReMAP process, the ratio of the average of the observed values of S to the average of the predicted values of S was used to calculate the bias correction factor, BCF.

7. 95% confidence intervals were calculated in terms of Ln(S) and intervals on Ln(S).

8. The predicted values of Ln(S) and the intervals on Ln(S) were converted to estimates of sigma and the confidence intervals on sigma through Exp(Ln(S)). The completed prediction equation is of the form Est Sigma = $BCF*kC^p$.

9. This equation... the line of regression... was traversed in small steps along the entire length of the line (from C = 500 to C = 5000. The upper and lower confidence limits were calculated at each step. The frequency with which those intervals encompassed the true values of Sigma (the true values of sigma calculated from Sigma = $0.2*C^{0.8}$) was recorded.

8. This entire process from Step 1 through Step 9 was repeated 10,000 times; in some instances 30,000 times.

The foregoing simulates getting 10,000 complete sets of ReMAP data; fitting those data to the model Ln(S) = Ln(K) + p*Ln(C); applying the appropriate bias correction factors; and (because we know the true value of Sigma in these simulations) observing how frequently the population of confidence intervals actually encompass the true values of sigma.

The simulations also provided information about biases in estimating sigma using the ReMAP process. The bias... 100*(Est Sigma - Sigma)/Sigma... ranged from -0.1% at low levels of C (hence low values of sigma) to +1.1% at high levels of C (high levels of sigma). This means that when the estimated value of sigma is, for example, 10, the true value of sigma could be as low as 9.89. Using the same example at a low level of concentration, the true value of sigma could be 10.01. This bias is trivial when compared to other considerations such as the width of the confidence intervals on sigma.

To verify the software used for these simulations, an identical copy of this software was written with the following changes. Instead of generating data from the model Sigma = $0.2^*C^{0.8}$, a model of the form $Y = 60 + 0.3^*X$ was used to generate the primary data. Also, instead of fitting the data to the model $Ln(S) = Ln(K) + p^*Ln(C)$... these data were fit to a model of the form $Y = a + bX$ where Y is simply the observed value of Y. No bias correction factors or transformations are required for this model. The usual 95% confidence intervals were calculated on the true mean value of Y in the manner prescribed in textbooks. With "95% confidence intervals" we expect 95% of those intervals to encompass the true mean values of Y. The simulations in this "check" software demonstrated that the 95% confidence intervals encompass the true mean (the true line of regression) 95% of the time, precisely as expected. (30,000 lines simulated). This verified the "check" software was performing as it was designed to perform; and therefore confirmed that the companion software for simulating the ReMAP process was designed properly.

When simulating the ReMAP process with the software designed for that purpose, between 97% and 98% of the "95% confidence intervals" encompassed the true values of sigma. So about 97% to 98% of the "95% confidence intervals" actually capture sigma. It was observed that the frequency of capturing sigma was slightly sensitive to the amount of data; simulations were run with 20, 40, 60, and 80 simultaneous samples.

97% - 98% is higher than the anticipated 95%. This means that these confidence intervals are a little broader than expected; so they capture or cover sigma more frequently than expected. In a sense this is good, because (when using calculated 95% intervals) we can state that "about 97% of the intervals calculated in this manner (*this manner* means the ReMAP process) will capture sigma. On the other hand, this increase in the frequency of capture is present because of some inefficiency in the ReMAP process.

Simulations were run to understand the nature of this inefficiency. The magnitude of this increase in capture frequency (97% - 98% as compared to 95%) depends upon the number of simultaneous samples. When we have two simultaneous samples, our "95% intervals" are actually 97% - 98% intervals. When we have five simultaneous samples, our "95% intervals" are actually 96% intervals. Very small samples (N = 2) produce a distribution of S that is highly non-normal. Samples of N = 5 will produce a distribution of S that is closer to normal. This is a primary cause of intervals that are somewhat broader than expected.

No real harm has been done by this small deviation from expected. It just means that we should remember that what we are calling 95% intervals are closer to 97% - 98% intervals. Our probability of capturing sigma with these intervals is a little higher than anticipated.

In the foregoing discussion we have described the performance of the ReMAP process in terms of *individual confidence intervals.* These intervals are calculated and presented throughout this report. This means that when we estimate a confidence interval at a point along the scale of C, we can declare that "95% of the intervals calculated in this manner will capture the true value of sigma. (We know it's really 97% - 98%, but for the purpose of this discussion we'll skip that detail.) This also means that (at this point on the scale of C) "the probability that *this* confidence interval has captured sigma is 95%". So there is a probability of about 5% (actually 2% - 3%) that sigma is a little larger than the calculated upper limit, or a little smaller than the calculated lower limit. In reality this reduces to a probability of "about 1% in each tail."

There is another concept that should be considered. This is the concept of a confidence interval on a line as a whole. These intervals are concerned with the probability that a confidence interval *captures the true values of sigma at every point along a line.* (over the range of the data). This can be stated as "the probability is X% that this confidence interval *has captured the true values of sigma at all points along the scale of concentration.* This is also the probability that there are *no individual confidence intervals* that fail to capture sigma. In this context, our attention is directed to the frequency at which sigma is captured from one end of the line to the other, rather than at individual points along a line.* Thus the frequency of capturing is in terms of "lines captured", not in terms of individual points captured. Capturing the entire line (the line is Sigma = $0.2 * C^{0.8}$) as opposed to individual points on the line is a rather stringent requirement.

---

* In this report the cited confidence intervals are in terms of individual intervals, or points on the line, not whole line intervals. Thus the statement "95% of the intervals calculated in this manner will encompass sigma" means 95% of the individual intervals. This is the usual interpretation of confidence intervals.

For more information concerning whole line confidence intervals, see Natrella, M. G.; Experimental Statistics; National Bureau of Standards Handbook 91; August 1963; pages 5-15 and 5-16.,

Simulations with the model $Y = 60 + 0.3{*}X$ (with the usual confidence intervals on the true mean of Y) show that when we are reporting 95% individual confidence intervals, this is equivalent to 86% confidence intervals on the whole line. Thus the probability is 86% that our 95% confidence intervals captured the true values of the mean of Y at every point along the entire line.

Simulations of the ReMAP process in its entirety with the model Sigma = $0.2{*}C^{0.8}$ (simultaneous samples of N = 2) show that the frequency of capture for the whole line confidence interval is 91% - 92%. This increase from 86% to 91% - 92% is due to the fact that individual intervals capture sigma at the rate of 97% - 98%, not 95%. Thus in the context of the ReMAP process, "the probability is about 91% - 92% that a specific confidence interval has captured the true values of sigma along the entire length of the line." Which means that about 8 - 9% of the whole line intervals fail to capture sigma *at some point along the line.* This could be a single individual confidence interval positioned at one point, or it could be a sequence of individual intervals.

Remember, in the context of a whole line confidence interval, a single failure to capture at one point along the line would be counted as a "failure to capture the entire line". This is a stringent requirement, indeed. Theory predicts, and detailed simulations confirm, that when an isolated "failure to capture" occurs (and this will happen, by definition, because approximately 2% - 3% of the individual confidence intervals will "fail to capture"), those "failures" will usually be such that the true value of sigma is just outside of the confidence intervals. This is especially true when we are dealing with the ReMAP model because its frequency of inclusion for individual intervals is 97% - 98%.

The precise outcomes from these simulations are dependent on several factors:

1. The constants k and p in the underlying model Sigma = kC^p
2. The amount of data in the simultaneous samples.
3. The number of samples used to build the models.
4. The points at which predictions are made on the scale of C.

This proprietary simulation software can be modified to study specific situations in more detail.

What If There Is No Relationship Between Est σ and Concentration?

If the t-ratio associated with p is trivial (notably less than 2) then we have failed to detect a relationship between Est σ and C. *This does not mean there is no relationship*; it only means that whatever relationship there may be, it was not detected in this set of data.

Our ability to detect such a relationship is influenced by the range over which the concentration was varied. If the data span a narrow range (viz., only about one decade, or a factor of 10) the dispersion in the values of S may be too small to detect the actual change in σ.

If the data are badly distributed along the scale of C... for instance, concentrated near the centroid instead of near the ends of the scale... our ability to establish the relationship between σ and C may be degraded. Thus our inability to establish a relationship may simply be due to data that are poorly distributed on the scale of concentration.

The inherent variation among individual values of S, cannot be overcome unless we have sufficient data. Attempting to establish a relationship between σ and C with insufficient data spread over narrow ranges virtually insures that the t-ratio for p will be low; so low, in fact, that we may not detect the presence of a relationship much less establish the nature of it.

Any of the foregoing factors, or combinations of those factors, can be responsible for a failure to detect a relationship between σ and C. It cannot be overemphasized that a "failure to detect" does not imply there is no relationship. Rather, it implies that we did not have sufficient data spread over a wide range of concentrations.

If the t-ratio for p is low (notably smaller than 2) then we may decide to take the following position. Since we have not established a relationship, then we may declare that σ is a constant. This is equivalent to declaring that the individual values of S effectively came from one common source whose true standard deviation is σ. Under this practice we could... if we elect to do so.... simply pool the individual values of S and use that as the estimate of σ. When doing this, it is appropriate to use the factors in Table 1 to calculate confidence bounds on σ.

The ITEQ data (Table 12 of the main body of this report) will be used to illustrate this practice. The individual standard deviations S are reported as 0.0457, 0.0012, ... , 0.0016. There are 22 pairs of N = 2; each of these provides a 1 df estimate of σ. To pool, square each value of S, sum and divide by 22 to obtain the pooled variance. The pooled Variance = 0.000712. Take the square root to find the pooled standard deviation; pooled S = 0.0267 with 22 df. Multiply 0.0267 by 0.777 and by 1.40 (interpolated in Table 1) to find the 95% confidence bounds on σ; 0.0207 and 0.0374. *No bias correction factor is required.** This is equivalent to drawing three horizontal lines on Fig. 23. These alternative limits are relatively narrow because they are not burdened with the uncertainly in estimating the slope, p.

---

* Pooling is not the same as averaging standard deviations. The small-sample bias correction factor is not needed when pooling if the number of pooled degrees of freedom is greater than 10.

Recommendations

Although the variation in simultaneously sampled data causes uncertainly in the relationship between σ and C, the width of the confidence intervals on σ can be minimized by acquiring data at well-chosen points along the scale of C. In planning for future data... anticipating a model of the form Est $\sigma = kC^p$... about one-half of the data should be collected at a "low" level and one-half should be collected at a "high" level of each pollutant. This will minimize SE(coeff), maximize the t-ratio on the power coefficient, and minimize the width of the confidence intervals on σ for a given amount of data.

If there is interest in pursuing alternative models, then the following rules apply. For every coefficient in the model used to associate σ with C, data should be concentrated at a point along the scale of C. This means that for a model with two estimated coefficients (viz., Est $\sigma = kC^p$) the experimental data should be concentrated at two points* on the scale of C, as in Table 5 and Fig. 5. For a model of the form Est $\sigma = a + kC^p$ the data should be concentrated at three distinct points* on the scale of C. If the data are poorly dispersed, as in Table 2 and Fig. 2 then the SE(Coeffs) will be large and the confidence intervals on estimates of Est σ may be wide. Attempting to estimate three model coefficients from poorly dispersed data can only lead to confusion.

There is no justification for using models more complex than Est $\sigma = kC^p$ with the ReMAP data at this time.

Within this report there is sufficient information to allow us to estimate the amount of data needed and the best positioning of that data on the scale of C so as to reduce the confidence intervals on σ to pre-specified widths. Although these calculations could be done analytically, it is better to do them with Monte Carlo simulations because of the complexity introduced by the bias corrections. Moreover, with simulations we can quickly explore "what if cases" before investing in additional data. A further advantage of Monte Carlo is that it makes the underlying models (and the assumptions) completely visible and unambiguous as compared to analytical methods that require in-depth knowledge of statistical methods.

---

* Data should be concentrated at additional points on the scale of C in order to test for lack-of-fit.